SIGNIFICANT CHANGES TO THE

INTERNATIONAL ENERGY CONSERVATION CODE®

2021 EDITION

SIGNIFICANT CHANGES TO THE
International Energy Conservation Code®

2021 EDITION

International Code Council

ICC Staff:

Executive Vice President and Director of Business Development:
Mark A. Johnson

Senior Vice President, Business and Product Development:
Hamid Naderi

Vice President and Technical Director, Products and Services:
Doug Thornburg

Senior Marketing Specialist:
Dianna Hallmark

ISBN: 978-1-952468-22-3 (print)
ISBN: 978-1-955052-47-4 (PDF download)

Project Head: Doug Thornburg

Publications Manager: Anne F. Kerr

Cover Design: Ricky Razo

INTERNATIONAL CODE COUNCIL®

Errata on various ICC publications may be available at www.iccsafe.org/errata.

First Printing: September 2021

PRINTED IN THE USA

T026706

Contents

Preface

The purpose of *Significant Changes to the International Energy Conservation Code®, 2021 Edition* is to familiarize building officials, plans examiners, inspectors, design professionals, contractors and others in the construction industry with many of the important changes in the 2021 *International Energy Conservation Code®* (IECC®). This publication is designed to assist those code users in identifying the specific code changes that have occurred and, more importantly, understanding the reasons behind the changes. It is also a valuable resource for jurisdictions in their code-adoption process.

Only a portion of the total number of code changes to the IECC are discussed in this book. The changes selected were identified for a number of reasons, including the frequency of application, special significance or change in application. However, the importance of those changes not included is not to be diminished. Further information on all code changes can be found in the *Complete Revision History to the 2021 I-Codes,* available from the International Code Council® (ICC®) online store. The revision history provides the published documentation for each successful code change contained in the 2021 IECC since the 2018 edition.

The IECC discussion in this publication addressing significant code changes is arranged to follow the general layout of the IECC, including code sections and section number format. The table of contents, in addition to providing guidance in the use of this publication, allows for the quick identification of the significant code changes that occur in the IECC.

Throughout the book, each change is accompanied by a photograph, an application example or an illustration to assist and enhance the reader's understanding of the specific change. A summary and a discussion of the significance of the changes are also provided. Each code change is identified by type, either an addition, modification, clarification or deletion.

The code change itself is presented in a format similar to the style utilized for code-change proposals. Deleted code language is shown with a strike-through, and new code text is indicated by underlining. As a result, the actual 2021 code language is provided as well as a comparison with the 2018 code language so the user can easily determine changes to the specific code text.

As with any code-change text, *Significant Changes to the International Energy Conservation Code, 2021 Edition* is best used as a study companion to the 2021 IECC. Because only a limited discussion of each change is provided, the code itself should always be referenced to gain a more comprehensive understanding of the code change and its application.

The commentary and opinions set forth in this text are those of the authors and do not necessarily represent the official position of the ICC. In addition, they may not represent the views of any enforcing agency, as such agencies have the sole authority to render interpretations of the IECC. In many cases, the explanatory material is derived from the reasoning expressed by the code-change proponent.

Comments concerning this publication are encouraged and may be directed to the ICC at significantchanges@iccsafe.org.

About the *International Energy Conservation Code*

The IECC is a comprehensive model code that regulates minimum energy-efficient provisions for new buildings and additions and alterations to existing buildings. The IECC contains two separate sets of provisions – commercial and residential. These provisions apply to the building thermal envelope and mechanical, service water heating and electrical power and lighting systems of applicable building types. Administrative provisions and definitions specific to each set of requirements are also included. The IECC integrates easy-to-understand prescriptive provisions for compliance as well as performance criteria that make possible the use of new and innovative materials, equipment and building designs.

The IECC is one of many codes in the family of International Codes published by the International Code Council (ICC). These codes are maintained and updated through an open code development process. Building officials, design professionals and others involved in the building construction industry recognize the need for a modern, up-to-date energy code addressing the design and installation of building systems through requirements emphasizing energy efficiency. The IECC, in the 2021 edition, is intended to meet these needs through model code regulations for energy efficient buildings in all communities, large and small.

The ICC was established in 1994 and is a nonprofit association that provides a wide range of building safety solutions including product evaluation, accreditation, certification, codification and training. The ICC develops model codes and standards used worldwide to construct safe, sustainable, affordable and resilient structures. The ICC's mission is to provide the highest-quality codes, standards, products and services for all concerned with the safety and performance of the built environment.

About the Authors

Michelle L. Britt, LEED AP
Partner, Britt/Makela Group

Michelle L. Britt was the Director of Energy Programs for the ICC from late 2017 through 2020. Prior to employment with ICC she spent 9 years as a partner with Britt/Makela Group as an energy consultant focused on energy codes, beyond code programs and energy efficiency domestically and in the Pacific Islands. Ms. Britt also spent 4 years with Pacific Northwest National Laboratory in the Building Energy Codes Program where she focused on deployment of the energy code and development of appliance standards. She has had the opportunity to work in the U.S. Territories of Guam, the Commonwealth of the Mariana Islands and American Samoa, Fiji, India and Saudi Arabia. Ms. Britt began her career as a planner with the town of Tiburon in California.

In her role with ICC, Ms. Britt was Secretariat for the IECC code development process and staff liaison to the Sustainability, Energy and High Performance Code Action Committee. She led ICC initiatives focused on IECC compliance, advanced energy efficiency and decarbonization, and supported energy efficiency internationally.

A graduate of Sonoma State University and the University of Idaho, Ms. Britt has 30 years of experience in the built environment. She has authored a wide variety of other code and beyond code support publications including *Evaluating Energy Code Compliance in Cities – Guidance and Methodology, Building America Best Practices Series Volume 13: Energy Performance Techniques and Technologies: Preserving Historic Homes* and *Going Beyond Code: A Guide to Creating Effective Green Building Programs for Energy Efficient and Sustainable Communities.*

Jerica Stacey
Energy Code Specialist, International Code Council

Jerica Stacey has over a decade of experience in the development, adoption and implementation of building energy codes and standards, as well as with energy code compliance evaluations and field studies and energy efficiency program evaluations. Ms. Stacey has worked in both the public and private sectors, supporting the energy codes and standards practice area at Cadmus, working as an energy code consultant with the Britt/Makela Group and contributing to projects under the U.S. Department of Energy's Building Energy Codes Program while at the Pacific Northwest National Laboratory. Ms. Stacey is currently an energy code specialist supporting ICC's Business and Product Development group.

Ms. Stacey has authored a wide variety of energy code support publications and studies, including the 2021 *Energy Code Essentials; 2018 CARICOM Energy Code Essentials; Attributing Codes and Standards Savings to Program Administrator Activities: Review of Approaches in Canada and the United States; Giving Credit Where Credit Is Due: Assessing Attribution and Savings from a Building Energy Code Compliance Enhancement Program; Compliance Verification Paths for Residential and Commercial Energy Codes*; and numerous impact and process evaluations for utilities, state government offices and program administrators.

About the International Code Council®

The International Code Council is a nonprofit association that provides a wide range of building safety solutions including product evaluation, accreditation, certification, codification and training. It develops model codes and standards used worldwide to construct safe, sustainable, affordable and resilient structures. ICC Evaluation Service (ICC-ES) is the industry leader in performing technical evaluations for code compliance, fostering safe and sustainable design and construction.

Washington DC Headquarters:
500 New Jersey Avenue, NW, 6th Floor
Washington, DC 20001

Regional Offices:
Eastern Regional Office: (BIR)
Central Regional Office: (CH)
Western Regional Office: (LA)

Distribution Center (Lenexa, KS)

1-888-ICC-SAFE (1-888-422-7233)
www.iccsafe.org

Family of Solutions:

ICC International Code Council®
ES ICC Evaluation Service®
IAS International Accreditation Service®
General Code
ICC SKGA S. K. Ghosh Associates®
NTA®
ICC Community Development Solutions
Alliance for National & Community Resilience®
ICC Pei

PART 1

Commercial Energy Provisions: Administration

Chapters 1 and 2

Photo courtesy of pawel.gaul/E+/Getty Images

- **Chapter 1** Scope and Administration
- **Chapter 2** Definitions

The administration chapter of the *International Energy Conservation Code* (IECC) covers the general scope, intent, applicability and enforcement of the IECC. Section C101.2 establishes the criteria for buildings that are regulated by the IECC Commercial provisions. Buildings outside the scope of Section C101.2 are regulated by the Residential provisions of the IECC. The administration provisions of Chapter 1 also cover the applicability of the code to mixed-use buildings; alternative materials, design and methods of construction; construction documents; inspection fees; referenced standards; and stop work orders.

Definitions used throughout the IECC are found in Chapter 2. The definitions contained within the IECC are intended to reflect the special meaning of such terms within the scope of the code. As terms can often have multiple meanings in their ordinary day-to-day use or within the various disciplines of the construction industry, it is important that their meanings be understood within the context of the IECC. ■

C103.1

Digital Plans

C103.2

Information on Construction Documents

C202

Definition of Biogas and Biomass

C202

Definitions of Fan Efficiencies

C103.1

Digital Plans

CHANGE TYPE: Clarification

CHANGE SUMMARY: Provisions related to energy code documentation clarify that construction documents and supporting data may be submitted digitally.

2021 CODE: C103.1 General. Construction documents and other supporting data shall be submitted in one or more sets, or in a digital format where allowed by the building official, with each application for a permit. The construction documents shall be prepared by a registered design professional where required by the statutes of the jurisdiction in which the project is to be constructed. Where special conditions exist, the code official is authorized to require necessary construction documents to be prepared by a registered design professional.

> **Exception:** The code official is authorized to waive the requirements for construction documents or other supporting data if the code official determines they are not necessary to confirm compliance with this code.

CHANGE SIGNIFICANCE: This provision was first drafted when blueprints were the standard, and construction plans and documentation were submitted on paper. As more jurisdictions move to an online, digital submittal process, the construction documents provisions proved confusing because the requirement for "one or more sets" implied to some that paper copies are still required. This code change clarifies that the code official may choose to accept digital submittals in lieu of paper documentation. The digital format facilitates a comprehensive electronic permitting, plan review and inspection system, and includes these benefits:

- Streamlined online submittals, payments, reviews and inspection scheduling
- Standardized building site information
- Readily available project status and information
- Real-time inspection results from the field.

Photo courtesy of SolStock

Plans may be submitted digitally.

C103.2

Information on Construction Documents

CHANGE TYPE: Modification

CHANGE SUMMARY: The information required on the construction documents has been expanded to include an indication of the energy compliance path used and additional air barrier details.

2021 CODE: C103.2 Information on construction documents. Construction documents shall be drawn to scale on suitable material. Electronic media documents are permitted to be submitted where approved by the code official. Construction documents shall be of sufficient clarity to indicate the location, nature and extent of the work proposed, and show in sufficient detail pertinent data and features of the building, systems and equipment as herein governed. Details shall include, but are not limited to, the following as applicable:

1. <u>Energy compliance path.</u>
2. Insulation materials and their *R*-values.
3. Fenestration *U*-factors and solar heat gain coefficients (SHGCs).
4. Area-weighted *U*-factor and solar heat gain coefficient (SHGC) calculations.
5. Mechanical system design criteria.
6. Mechanical and service water-heating systems and equipment types, sizes and efficiencies.
7. Economizer description.
8. Equipment and system controls.
9. Fan motor horsepower (hp) and controls.
10. Duct sealing, duct and pipe insulation and location.
11. Lighting fixture schedule with wattage and control narrative.
12. Location of daylight zones on floor plans.
13. Air ~~sealing details~~.<u>barrier and air sealing details, including the location of the air barrier.</u>

CHANGE SIGNIFICANCE: Two additions were made to the information required on construction documents to aid in plan review and inspection. The first is the identification of the compliance path. This information is essential for the plan reviewer to determine if the project demonstrates compliance with the specific energy requirements. When this information is not provided on the construction documents and plans examiners are required to investigate, the review process is lengthened. This extension may be in the form of a review comment or the time it takes to retrieve the information by email or phone. Providing the energy code compliance path details at submittal allows the plans examiner to review the plans according to the intended energy compliance path the architect/designer chose for the project.

Additionally, the requirement for air sealing documentation is clarified, providing all parties involved a clear understanding of what is required and, thus, what is to be reviewed and inspected.

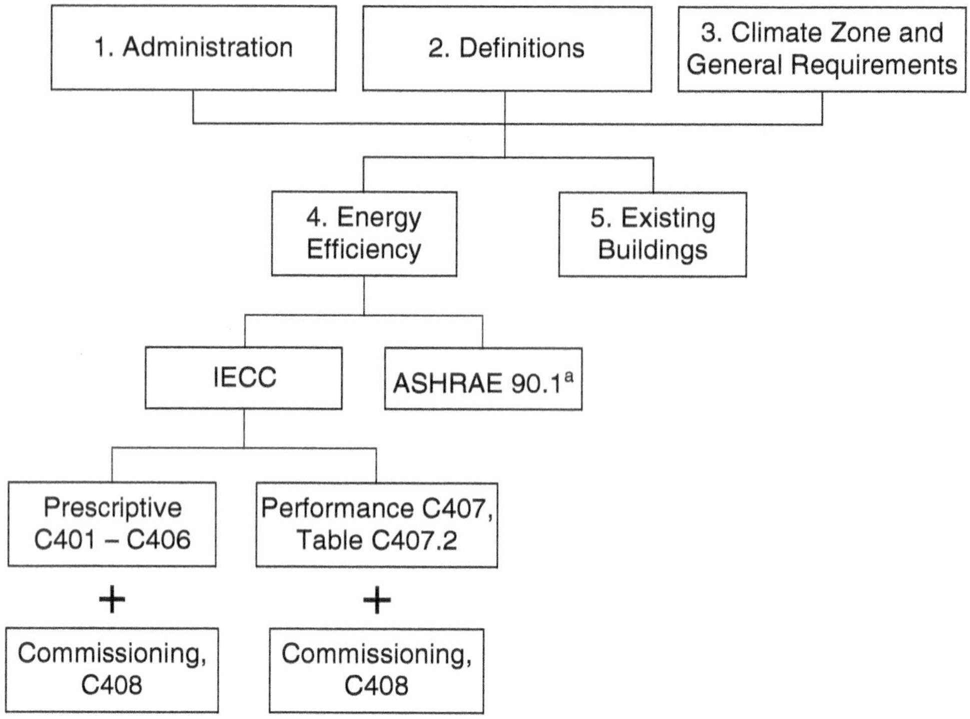

[a] ANSI/ASHRAE/IES Standard 90.1-2019

Energy compliance path must be noted on construction documents.

C202

Definition of Biogas and Biomass

CHANGE TYPE: Addition

CHANGE SUMMARY: Definitions for biogas, biomass and renewable energy resources are introduced and the definition of on-site renewable energy is clarified.

2021 CODE: **BIOGAS.** A mixture of hydrocarbons that is a gas at 60°F (15.5°C) and 1 atmosphere of pressure that is produced through the anaerobic digestion of organic matter.

BIOMASS. Nonfossilized and biodegradable organic material originating from plants, animals and/or micro-organisms, including products, by-products, residues and waste from agriculture, forestry and related industries as well as the nonfossilized and biodegradable organic fractions of industrial and municipal wastes, including gases and liquids recovered from the decomposition of nonfossilized and biodegradable organic material.

ON-SITE RENEWABLE ENERGY. Energy ~~derived~~ from renewable energy resources harvested at the building project site. ~~solar radiation, wind, waves, tides, landfill gas, biogas, biomass or the internal heat of the earth. The energy system providing onsite renewable energy shall be located on the project site.~~

RENEWABLE ENERGY RESOURCES. Energy derived from solar radiation, wind, waves, tides, landfill gas, biogas, biomass or extracted from hot fluid or steam heated within the earth.

CHANGE SIGNIFICANCE: The definition of on-site renewable energy was first introduced in the 2012 IECC. When it was written in 2010, it was the first time that renewable energy had been defined in an I-code,

Photo courtesy of kontrast-fotodesign/iStock

Biomass is a form of renewable energy.

and it reflected a very early understanding of the industry. The definition had not been significantly revised until the 2021 IECC clarified that the resources must be harvested at the project site to be considered on-site renewable energy.

A new definition for renewable energy resources delineates exactly what constitutes a renewable energy resource and includes both biogas and biomass. The definition for biomass sources is limited to those that meet specifications as waste products, as defined in the glossary of the Energy Information Administration. There are many types of biomass energy and this definition ensures that virgin material of unknown origin is not used as a steady source of energy. Collectively, these new and revised definitions support the provisions in Section C406 for on-site renewable energy, which are covered in Part 3 of this text.

C202

Definitions of Fan Efficiencies

CHANGE TYPE: Addition

CHANGE SUMMARY: New definitions related to fans introduce and support the Fan Energy Index, which replaces the Fan Efficiency Grade metric and aligns the IECC more closely with the fan requirements in ASHRAE 90.1.

2021 CODE: **FAN, EMBEDDED.** A fan that is part of a manufactured assembly where the assembly includes functions other than air movement.

FAN ARRAY. Multiple fans in parallel between two plenum sections in an air distribution system.

FAN ENERGY INDEX (FEI). The ratio of the electric input power of a reference fan to the electric input power of the actual fan as calculated in accordance with AMCA 208.

FAN NAMEPLATE ELECTRICAL INPUT POWER. The nominal electrical input power rating stamped on a fan assembly nameplate.

FAN SYSTEM ELECTRICAL INPUT POWER. The sum of the fan electrical power of all fans that are required to operate at fan system design conditions to supply air from the heating or cooling source to the conditioned spaces and/or return it to the source or exhaust it to the outdoors.

Photo courtesy of AMCA international, Nordroden/ bigstock.com

The Fan Energy Index replaces the Fan Efficiency Grade.

CHANGE SIGNIFICANCE: The Fan Energy Index (FEI) is defined as a ratio of the electrical input power of a reference fan (FEPref) to the electrical input power of the actual fan (FEPact) for which the FEI is calculated. The reference fan concept is used to normalize the FEI calculation to a consistent power level independent of fan type, fan drive components or any regulatory requirements. FEPact is the fan electrical input power associated with a given fan duty point in terms of airflow and pressure. There are slightly different FEI equations for fans rated using total pressure and those rated using static pressure. An FEI value less than 1.0 indicates that the fan does not meet the efficiency level being analyzed, while a value greater than 1.0 indicates that the fan is more efficient than the efficiency level being analyzed.

Replacing the Fan Efficiency Grade (FEG) metric with the FEI results in a more effective energy savings metric. The FEI was developed in response to the U.S. Department of Energy (DOE) rulemaking for commercial fans and blowers in which a wire-to-air metric was deemed to be more effective at saving energy because it would consider the impacts of motors and drives on fan energy performance.

The use of the FEI also aligns with updates in ASHRAE 90.1-2019, with the significant difference being that the IECC requires FEI ratings be determined in accordance with AMCA 208 by an approved, independent testing laboratory and labeled by the manufacturer.

Photo courtesy of AzmanL/E+/Getty Images

PART 2

Commercial Energy Provisions: General Requirements

Chapter 3

- **Chapter 3** General Requirements

C301.3

Climate Zone Definitions

Chapter 3 provides the foundational requirements for the provisions in Chapters 4 and 5.

Climate zone designations in Chapter 3 are used as a reference for the building envelope and mechanical provisions that are climate-specific. The code establishes many requirements, such as wall and roof insulation *R*-values and window and door thermal transmittance (*U*-factors), that affect the mechanical systems based on the climate where the building is located.

In addition, Chapter 3 provides interior design conditions that are used as a basis for assumptions in heating and cooling load calculations and provides the basic material requirements for insulation and fenestration. The requirements for labeling and product rating are also provided; these are foundational to the envelope requirements of Chapter 4. Finally, Chapter 3 provides specifications for insulation installation that help ensure the effectiveness of the insulation. ■

C301.3

Climate Zone Definitions

CHANGE TYPE: Modification

CHANGE SUMMARY: The climate zone provisions have been comprehensively updated and include a new Climate Zone 0. Ten percent of U.S. counties are now in a new climate zone.

2021 CODE TEXT: **C301.3 ~~International climate zones~~ Climate zone definitions.** ~~The climate zone for any location outside the United States shall be determined by applying Table C301.3(1) and then Table C301.3(2).~~ To determine the climate zones for locations not listed in this code, use the following information to determine climate zone numbers and letters in accordance with Items 1 through 5.

1. Determine the thermal climate zone, 0 through 8, from Table C301.3 using the heating (HDD) and cooling degree-days (CDD) for the location.
2. Determine the moisture zone (Marine, Dry or Humid) in accordance with Items 2.1 through 2.3.
   - 2.1. If monthly average temperature and precipitation data are available, use the Marine, Dry, and Humid definitions to determine the moisture zone (C, B or A).
   - 2.2. If annual average temperature information (including degree-days) and annual precipitation (i.e., annual mean) are available, use Items 2.2.1 through 2.2.3 to determine the moisture zone. If the moisture zone is not Marine, then use the Dry definition below to determine whether Dry or Humid.
     - 2.2.1. If thermal climate zone is 3 and CDD50°F ≤ 4,500 (CDD10°C ≤2500), climate zone is Marine (3C).
     - 2.2.2. If thermal climate zone is 4 and CDD50°F ≤ 2,700 (CDD10°C ≤ 1500), climate zone is Marine (4C).
     - 2.2.3. If thermal climate zone is 5 and CDD50°F ≤ 1,800 (CDD10°C ≤ 1000), climate zone is Marine (5C).

Photo courtesy of Daniel Schnherr / EyeEm/Getty Images

Photo courtesy of kolderal/Moment/ Getty Images

The IECC provisions are climate-specific.

2.3. If only degree-day information is available, use Items 2.3.1 through 2.3.3 to determine the moisture zone. If the moisture zone is not Marine, then it is not possible to assign Humid or Dry moisture zone for this location.

2.3.1. If thermal climate zone is 3 and CDD50°F ≤ 4,500 (CDD10°C ≤ 2500), climate zone is Marine (3C).

2.3.2. If thermal climate zone is 4 and CDD50°F ≤ 2,700 (CDD10°C ≤ 1500), climate zone is Marine (4C).

2.3.3. If thermal climate zone is 5 and CDD50°F ≤ 1,800 (CDD10°C ≤ 1000), climate zone is Marine (5C).

3. Marine (C) Zone definition: Locations meeting all the criteria in Items 3.1 through 3.4.

3.1. Mean temperature of coldest month between 27°F (–3°C) and 65°F (18°C).

3.2. Warmest month mean < 72°F (22°C).

3.3. Not fewer than four months with mean temperatures over 50°F (10°C).

3.4. Dry season in summer. The month with the heaviest precipitation in the cold season has at least three times as much precipitation as the month with the least precipitation in the rest of the year. The cold season is October through March in the Northern Hemisphere and April through September in the Southern Hemisphere.

4. Dry (B) definition: Locations meeting the criteria in Items 4.1 through 4.4.

4.1. Not Marine (C).

4.2. If 70 percent or more of the precipitation, P, occurs during the high sun period, defined as April through September in the Northern Hemisphere and October through March in the Southern Hemisphere, then the dry/humid threshold is in accordance with Equation 3-1.

$$P < 0.44 \times (T - 7)$$

$$[P < 20.0 \times (T + 14) \text{ in SI units}] \quad \text{(Equation 3-1)}$$

where:

P = Annual precipitation, inches (mm).
T = Annual mean temperature, °F (°C).

4.3. If between 30 and 70 percent of the precipitation, P, occurs during the high sun period, defined as April through September in the Northern Hemisphere and October through March in the Southern Hemisphere, then the dry/humid threshold is in accordance with Equation 3-2.

$$P < 0.44 \times (T - 19.5)$$

$$[P < 20.0 \times (T + 7) \text{ in SI units}] \quad \text{(Equation 3-2)}$$

where:

P = Annual precipitation, inches (mm).

T = Annual mean temperature, °F (°C).

4.4. If 30 percent or less of the precipitation, P, occurs during the high sun period, defined as April through September in the Northern Hemisphere and October through March in the Southern Hemisphere, then the dry/humid threshold is in accordance with Equation 3-3.

$P < 0.44 \times (T - 32)$

$[P < 20 \times T \text{ in SI units}]$ (Equation 3-3)

where:

P = Annual precipitation, inches (mm).

T = Annual mean temperature, °F (°C).

5. Humid (A) definition: Locations that are not Marine (C) or Dry (B).

TABLE C301.1 Climate Zones, Moisture Regimes, and Warm Humid Designations by State, County and Territory[a]

US STATES

ALABAMA
~~3A~~ 2A Coffee*
~~3A~~ 2A Covington*
~~3A~~ 2A Dale*
~~3A~~ 2A Escambia*
~~3A~~ 2A Geneva*
~~3A~~ 2A Henry*
~~3A~~ 2A Houston*

ALASKA
~~8~~ 7 Bethel
~~7~~ 8 Denali
~~8~~ 7 Dillingham
~~7~~ 6A Haines
~~7~~ 6A Juneau
~~7~~ 5C Ketchikan Gateway
~~7~~ 6A Kodiak Island
~~7~~ 5C Prince of Wales Outer Ketchikan
~~7~~ 5C Sitka
~~7~~ 6A Skagway-Hoonah-Angoon
~~7~~ 6A Wrangell-Petersburg

ARIZONA

ARKANSAS

CALIFORNIA

COLORADO
~~5B~~ 4B Bent
~~6B~~ 5B Custer
~~5B~~ 4B Prowers

CONNECTICUT

DELAWARE

DISTRICT OF COLUMBIA

FLORIDA
~~2A~~ 1A Palm Beach*

GEORGIA
~~4A~~ 3A Banks
~~3A~~ 2A Calhoun*
~~4A~~ 3A Catoosa
~~4A~~ 3A Chattooga
~~3A~~ 2A Coffee*
~~4A~~ 3A Dade
~~4A~~ 3A Dawson
~~3A~~ 2A Dougherty*
~~3A~~ 2A Early*
~~4A~~ 3A Fannin
~~4A~~ 3A Floyd
~~4A~~ 3A Franklin
~~4A~~ 3A Gilmer
~~4A~~ 3A Gordon
~~4A~~ 3A Habersham
~~4A~~ 3A Hall
~~4A~~ 3A Lumpkin
~~4A~~ 3A Murray
~~4A~~ 3A Pickens
~~4A~~ 3A Rabun
~~4A~~ 3A Stephens
~~3A~~ 2A Tift*
~~4A~~ 3A Towns
~~4A~~ 3A Union
~~4A~~ 3A Walker
~~4A~~ 3A White
~~4A~~ 3A Whitfield
~~3A~~ 2A Worth*

HAWAII

IDAHO

ILLINOIS
~~5A~~ 4A Calhoun
~~5A~~ 4A Clark
~~5A~~ 4A Coles
~~5A~~ 4A Cumberland
~~5A~~ 4A Greene
~~5A~~ 4A Jersey

INDIANA
~~5A~~ 4A Bartholomew
~~5A~~ 4A Clay
~~5A~~ 4A Decatur
~~5A~~ 4A Fayette
~~5A~~ 4A Franklin
~~5A~~ 4A Hendricks
~~5A~~ 4A Johnson
~~5A~~ 4A Marion
~~5A~~ 4A Morgan
~~5A~~ 4A Owen
~~5A~~ 4A Putnam
~~5A~~ 4A Rush
~~5A~~ 4A Shelby
~~5A~~ 4A Union
~~5A~~ 4A Vigo

IOWA
~~6A~~ 5A Allamakee
~~6A~~ 5A Bremer
~~6A~~ 5A Buchanan
~~6A~~ 5A Buena Vista
~~6A~~ 5A Butler
~~6A~~ 5A Calhoun
~~6A~~ 5A Cherokee
~~6A~~ 5A Chickasaw
~~6A~~ 5A Clayton
~~6A~~ 5A Delaware
~~6A~~ 5A Fayette
~~6A~~ 5A Floyd
~~6A~~ 5A Franklin
~~6A~~ 5A Grundy
~~6A~~ 5A Hamilton
~~6A~~ 5A Hardin
~~6A~~ 5A Howard
~~6A~~ 5A Humboldt
~~6A~~ 5A Ida
~~6A~~ 5A Plymouth
~~6A~~ 5A Pocahontas
~~6A~~ 5A Sac
~~6A~~ 5A Webster
~~6A~~ 5A Winneshiek
~~6A~~ 5A Wright

KANSAS
~~5A~~ 4A Cloud
~~5A~~ 4A Ellis
~~5A~~ 4A Graham
~~5A~~ 4A Hamilton
~~5A~~ 4A Lane
~~5A~~ 4A Mitchell
~~5A~~ 4A Ness
~~5A~~ 4A Osborne
~~5A~~ 4A Rooks
~~5A~~ 4A Trego

(*continued*)

TABLE C301.1 *(continued)*

KENTUCKY

LOUISIANA

MAINE

MARYLAND
~~4A~~ 5A Allegany

MASSACHUSETTS

MICHIGAN
~~7~~ 6A Baraga
~~7~~ 6A Chippewa
~~7~~ 6A Gogebic
~~7~~ 6A Houghton
~~6A~~ 5A Huron
~~7~~ 6A Iron
~~7~~ 6A Luce
~~7~~ 6A Mackinac
~~6A~~ 7 Marquette
~~7~~ 6A Ontonagon
~~6A~~ 5A Sanilac
~~7~~ 6A Schoolcraft

MINNESOTA
~~7~~ 6A Becker
~~7~~ 6A Clay
~~6A~~ 5A Fillmore
~~7~~ 6A Grant
~~6A~~ 5A Houston
~~7~~ 6A Kanabec
~~7~~ 6A Mille Lacs
~~7~~ 6A Otter Tail
~~7~~ 6A Wilkin
~~6A~~ 5A Winona

MISSISSIPPI
~~3A~~ 2A George*

MISSOURI
~~5A~~ 4A Buchanan
~~5A~~ 4A Caldwell
~~5A~~ 4A Chariton
~~5A~~ 4A Clinton
~~4A~~ 3A Dunklin
~~4A~~ 3A Pemiscot

MONTANA

NEBRASKA

NEVADA
~~5B~~ 4B Carson City (city)
~~5B~~ 4B Douglas
~~5B~~ 4B Esmeralda
~~5B~~ 4B Lincoln
~~5B~~ 4B Lyon
~~5B~~ 4B Mineral
~~5B~~ 4B Nye

NEW HAMPSHIRE
~~6A~~ 5A Merrimack

NEW JERSEY
~~5A~~ 4A Mercer

NEW MEXICO
~~5B~~ 4B Catron
~~4B~~ 3B Sierra

NEW YORK
~~6A~~ 5A Allegany
~~6A~~ 5A Broome
~~6A~~ 5A Cattaraugus
~~6A~~ 5A Schoharie
~~6A~~ 5A Schuyler
~~6A~~ 5A Steuben
~~6A~~ 5A Tompkins
~~6A~~ 5A Wyoming

NORTH CAROLINA
~~4A~~ 3A Alamance
~~4A~~ 3A Alexander
~~4A~~ 3A Bertie
~~4A~~ 3A Caswell
~~4A~~ 3A Catawba
~~4A~~ 3A Chatham
~~4A~~ 3A Cherokee
~~4A~~ 3A Clay
~~4A~~ 3A Cleveland
~~4A~~ 3A Davie
~~4A~~ 3A Durham
~~4A~~ 3A Forsyth
~~4A~~ 3A Franklin
~~4A~~ 3A Gates
~~4A~~ 3A Granville
~~4A~~ 3A Guilford
~~4A~~ 3A Halifax
~~4A~~ 3A Harnett
~~4A~~ 3A Hertford
~~4A~~ 3A Iredell
~~4A~~ 3A Lee
~~4A~~ 3A Lincoln
~~5A~~ 4A Mitchell
~~4A~~ 3A Nash
~~4A~~ 3A Northampton
~~4A~~ 3A Orange
~~4A~~ 3A Person
~~4A~~ 3A Polk
~~4A~~ 3A Rockingham
~~4A~~ 3A Rutherford
~~4A~~ 3A Vance
~~4A~~ 3A Wake
~~4A~~ 3A Warren
~~4A~~ 3A Wilkes

NORTH DAKOTA
~~7~~ 6A Barnes
~~7~~ 6A Cass
~~7~~ 6A Eddy
~~7~~ 6A Foster
~~7~~ 6A Griggs
~~7~~ 6A Kidder
~~7~~ 6A McLean
~~7~~ 6A Mountrail
~~7~~ 6A Sheridan
~~7~~ 6A Steele
~~7~~ 6A Stutsman
~~7~~ 6A Traill
~~7~~ 6A Wells
~~7~~ 6A Williams

OHIO
~~5A~~ 4A Athens
~~5A~~ 4A Butler
~~5A~~ 4A Clinton
~~5A~~ 4A Fayette
~~5A~~ 4A Franklin
~~5A~~ 4A Greene
~~5A~~ 4A Highland
~~5A~~ 4A Hocking
~~5A~~ 4A Jackson
~~5A~~ 4A Madison
~~5A~~ 4A Meigs
~~5A~~ 4A Pickaway
~~5A~~ 4A Ross
~~5A~~ 4A Vinton
~~5A~~ 4A Warren

OKLAHOMA
~~3A~~ 4A Alfalfa
~~3A~~ 4A Craig
~~3A~~ 4A Delaware
~~3A~~ 4A Ellis
~~3A~~ 4A Garfield
~~3A~~ 4A Grant
~~3A~~ 4A Harper
~~3A~~ 4A Kay
~~3A~~ 4A Major

(continued)

TABLE C301.1 *(continued)*

~~3A~~ 4A Nowata
~~3A~~ 4A Osage
~~3A~~ 4A Ottawa
~~3A~~ 4A Washington
~~3A~~ 4A Woods
~~3A~~ 4A Woodward

OREGON

PENNSYLVANIA
~~5A~~ 4A Adams
~~5A~~ 4A Berks
~~6A~~ 5A Cameron
~~6A~~ 5A Clearfield
~~5A~~ 4A Cumberland
~~5A~~ 4A Dauphin
~~6A~~ 5A Elk
~~5A~~ 4A Franklin
~~5A~~ 4A Lancaster
~~5A~~ 4A Lebanon
~~6A~~ 5A McKean
~~5A~~ 4A Perry
~~6A~~ 5A Potter
~~6A~~ 5A Susquehanna
~~6A~~ 5A Tioga
~~6A~~ 5A Wayne

RHODE ISLAND

SOUTH CAROLINA
~~3A~~ 2A Beaufort*
~~3A~~ 2A Jasper*

SOUTH DAKOTA
~~6A~~ 5A Brule
~~6A~~ 5A Haakon
~~6A~~ 5A Jones
~~6A~~ 5A Lyman
~~6A~~ 5A Stanley

TENNESSEE
~~4A~~ 3A Bedford
~~4A~~ 3A Coffee
~~4A~~ 3A Davidson
~~4A~~ 3A Decatur
~~4A~~ 3A Franklin
~~4A~~ 3A Gibson
~~4A~~ 3A Giles
~~4A~~ 3A Grundy
~~4A~~ 3A Hamilton
~~4A~~ 3A Hickman
~~3A~~ 4A Lake
~~4A~~ 3A Lawrence
~~4A~~ 3A Lewis
~~4A~~ 3A Lincoln
~~4A~~ 3A Marion
~~4A~~ 3A Marshall
~~4A~~ 3A Maury
~~4A~~ 3A Moore
~~4A~~ 3A Perry
~~4A~~ 3A Rutherford
~~4A~~ 3A Wayne
~~4A~~ 3A Williamson

TEXAS
~~2A~~ 1A Cameron*
~~3A~~ 2A Dallas*
~~3A~~ 2A Ellis*
~~2A~~ 1A Hidalgo*
~~3A~~ 2A Johnson*
~~3A~~ 2A Navarro*
~~3A~~ 2A Tarrant*
~~2A~~ 1A Willacy*

UTAH
~~6B~~ 5B Box Elder
~~6B~~ 5B Cache
~~6B~~ 5B Carbon

VERMONT

VIRGINIA
4A (all except as follows:)
5A Alleghany
5A Bath
3A Brunswick
3A Chesapeake
5A Clifton Forge
5A Covington
3A Emporia
3A Franklin
3A Greensville
3A Halifax
3A Hampton
5A Highland
3A Isle of Wight
3A Mecklenburg
3A Newport News
3A Norfolk
3A Pittsylvania
3A Portsmouth
3A South Boston
3A Southampton
3A Suffolk
3A Surry
3A Sussex
3A Virginia Beach

WASHINGTON
~~4C~~ 5C Clallam
~~4C~~ 5C Island
~~4C~~ 5C Kitsap
~~6B~~ 5B Okanogan
~~4C~~ 5C San Juan

WEST VIRGINIA
~~5A~~ 4A Doddridge
~~5A~~ 4A Fayette
~~5A~~ 4A Greenbrier
~~5A~~ 4A Lewis
~~5A~~ 4A Nicholas
~~5A~~ 4A Raleigh
~~5A~~ 4A Summers
~~5A~~ 4A Upshur
~~5A~~ 4A Webster

WISCONSIN
~~6A~~ 5A Adams
~~7~~ 6A Ashland
~~7~~ 6A Bayfield
~~7~~ 6A Burnett
~~6A~~ 5A Calumet
~~6A~~ 5A Columbia
~~6A~~ 5A Crawford
~~6A~~ 5A Dane
~~6A~~ 5A Dodge
~~7~~ 6A Douglas
~~7~~ 6A Florence
~~6A~~ 5A Fond du Lac
~~7~~ 6A Forest
~~6A~~ 5A Grant
~~6A~~ 5A Green
~~6A~~ 5A Green Lake
~~6A~~ 5A Iowa
~~7~~ 6A Iron
~~6A~~ 5A Jefferson
~~6A~~ 5A Juneau
~~6A~~ 5A Kenosha
~~6A~~ 5A La Crosse
~~6A~~ 5A Lafayette
~~7~~ 6A Langlade
~~7~~ 6A Lincoln
~~6A~~ 5A Milwaukee
~~6A~~ 5A Monroe
~~7~~ 6A Oneida

(continued)

TABLE C301.1 ***(continued)***

~~6A~~ 5A Outagamie
~~6A~~ 5A Ozaukee
~~7~~ 6A Price
~~6A~~ 5A Racine
~~6A~~ 5A Richland
~~6A~~ 5A Rock
~~6A~~ 5A Sauk
~~7~~ 6A Sawyer
~~7~~ 6A Taylor
~~6A~~ 5A Vernon
~~7~~ 6A Vilas
~~6A~~ 5A Walworth
~~7~~ 6A Washburn
~~6A~~ 5A Washington
~~6A~~ 5A Waukesha
~~6A~~ 5A Waushara
~~6A~~ 5A Winnebago

WYOMING
~~6B~~ 5B Laramie

US TERRITORIES
AMERICAN SAMOA

GUAM

NORTHERN MARIANA ISLANDS

PUERTO RICO
1A (all except as follows:)*
2B Barraquitas
2B Cayey

VIRGIN ISLANDS

a. Key: A – Moist, B – Dry, C – Marine. Absence of moisture designation indicates moisture regime is irrelevant.
Asterisk (*) indicates a Warm Humid location.
Note: Only counties assigned a new climate zone are shown here. Please refer to Table C301.1 for the full list of counties.

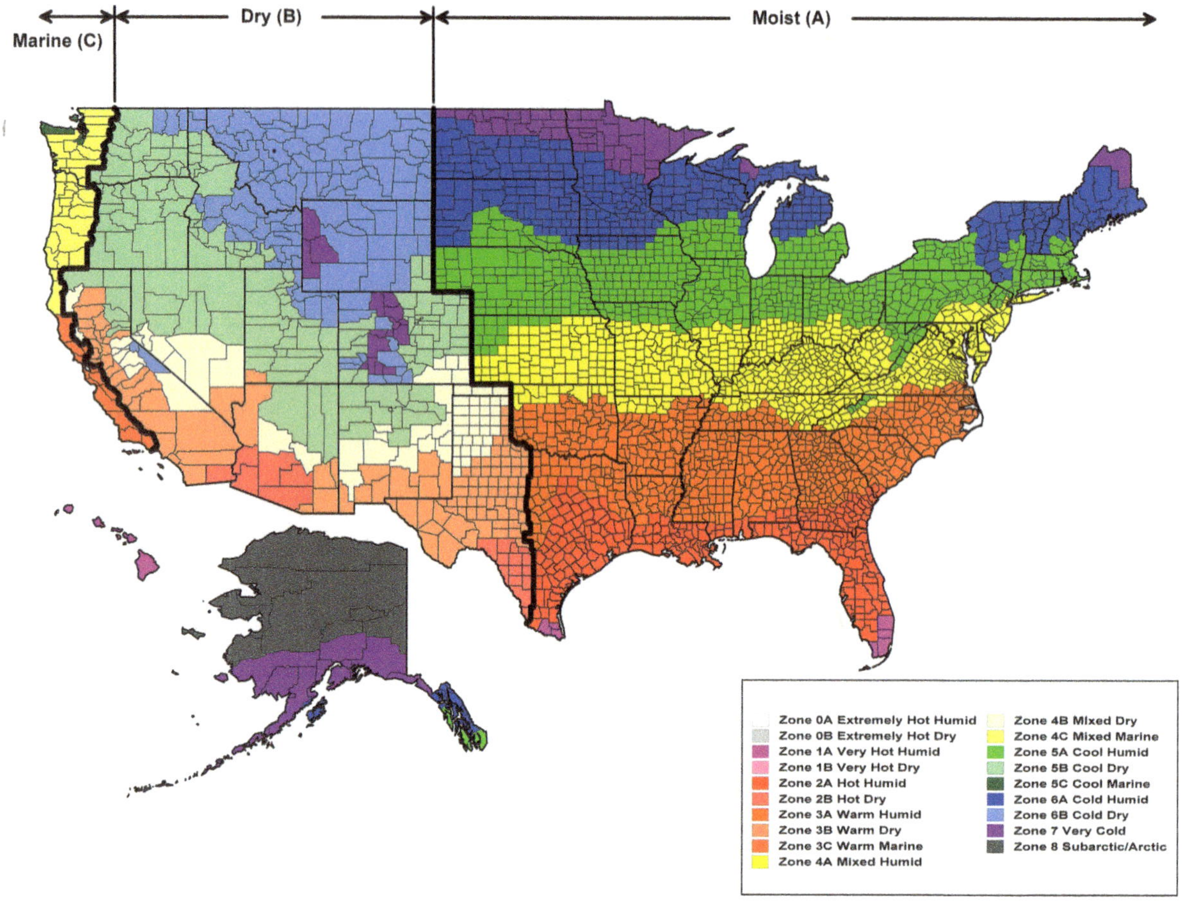

Figure C301.1 Climate Zones.

~~**TABLE C301.3(1) International Climate Zone Definitions**~~

~~MAJOR CLIMATE TYPE DEFINITIONS~~
~~**Marine (C) Definition—Locations meeting all four criteria:**~~ ~~1. Mean temperature of coldest month between −3°C (27°F) and 18°C (65°F).~~ ~~2. Warmest month mean < 22°C (72°F).~~ ~~3. At least four months with mean temperatures over 10°C (50°F).~~ ~~4. Dry season in summer. The month with the heaviest precipitation in the cold season has at least three times as much precipitation as the month with the least precipitation in the rest of the year. The cold season is October through March in the Northern Hemisphere and April through September in the Southern Hemisphere.~~
~~**Dry (B) Definition—Locations meeting the following criteria:**~~ ~~Not marine and $P_{in} < 0.44 \times (TF - 19.5)$ [$P_{cm} < 2.0 \times (TC + 7)$ in SI units]~~ ~~where:~~ ~~P_{in} = Annual precipitation in inches (cm)~~ ~~T = Annual mean temperature in °F (°C)~~
~~**Moist (A) Definition—Locations that are not marine and not dry.**~~
~~**Warm-humid Definition—Moist (A) locations where either of the following wet-bulb temperature conditions shall occur during the warmest**~~ ~~**six consecutive months of the year:**~~ ~~1. 67°F (19.4°C) or higher for 3,000 or more hours; or~~ ~~2. 73°F (22.8°C) or higher for 1,500 or more hours.~~

~~For SI: °C = [(°F) − 32]/1.8, 1 inch = 2.54 cm.~~

TABLE ~~C301.3(2)~~ C301.3 Thermal Climate Zone Definitions

ZONE NUMBER	THERMAL CRITERIA	
	IP Units	SI Units
0	10,800 < CDD50°F	6000 < CDD10°C
1	9,000 < CDD50°F ≤ 10,800	5000 < CDD10°C ≤ 6000
2	6,300 < CDD50°F ≤ 9,000	3500 < CDD10°C ≤ 5000
~~3A and 3B~~ 3	~~4,500 <~~ CDD50°F ≤ 6,300 AND HDD65°F ≤ ~~5,400~~ 3,600	CDD10°C < 3500 AND HDD18°C ≤ ~~3000~~ 2000
~~4A and 4B~~ 4	CDD50°F ≤ ~~4,500~~ 6,300 AND 3,600 < HDD65°F ≤ 5,400	CDD10°C < ~~2500~~ 3500 AND 2000 < HDD18°C ≤ 3000
~~3C~~	~~HDD65°F ≤ 3,600~~	~~HDD18°C ≤ 2000~~
~~4C~~	~~3,600 < HDD65°F ≤ 5,400~~	~~2000 < HDD18°C ≤ 3000~~
5	CDD50°F < 6,300 AND 5,400 < HDD65°F ≤ 7,200	CDD10°C < 3500 AND 3000 < HDD18°C ≤ 4000
6	7,200 < HDD65°F ≤ 9,000	4000 < HDD18°C ≤ 5000
7	9,000 < HDD65°F ≤ 12,600	5000 < HDD18°C ≤ 7000
8	12,600 < HDD65°F	7000 < HDD18°C

For SI: °C = [(°F)-32]/1.8.

CHANGE SIGNIFICANCE: The climate zone provisions in Chapter 3, from Section C301.3 through the climate zone tables and map, have undergone a significant update. This update corresponds with ASHRAE Standard 169-2013, which is referenced in the 2018 International Green Construction Code® (IgCC®), ANSI/ASHRAE/IES Standard 90.1 – Energy Standard for Buildings Except Low-Rise Residential Buildings (ASHRAE 90.1) and ANSI/ASHRAE/IES Standard 90.2 – Energy-Efficient Design of Low-Rise Residential Buildings (ASHRAE 90.2). ICC has a licensing agreement with ASHRAE to include the climate zone map, definitions and tables for consistency with ASHRAE Standard 169 and other key codes and standards.

These two key updates are relevant to most code users:

- Updates to the climate zone designations in Table C301.1 for approximately 10 percent of the counties in the United States. In general, the counties were assigned a warmer climate zone.
- Updates to Figure C301.1, the climate zone map, to reflect the changes in county climate zone designations.

The number of climate zones increased to include Climate Zone 0, representing the hottest of climates. Climate Zone 0 is a subset of the previous Climate Zone (CZ) 1, which included all locations with more than 9,000 Cooling Degree Days (CDDs). Climate Zone 1 now tops out at 10,800 CDDs, and Climate Zone 0 includes those locations with more than that. Cities in Climate Zone 0 include extremely hot locations such as Mumbai (Bombay), Jakarta and Abu Dhabi. There are no cities in the United States in Climate Zone 0; Miami and the islands of Hawaii are in Climate Zone 1. Several code changes introduce specific provisions for

Climate Zone 0. Because Climate Zone 0 is a subset of the former Climate Zone 1, most references throughout the IECC to CZ 1 have been edited to CZ 0 and 1.

Code users needing to calculate the moisture zone (marine, dry or humid) will find Section C301.3 entirely revised. Calculations are in three categories: (a) when monthly average temperature and precipitation data are available, (b) when annual average temperature, degree days and precipitation are available, and (c) when only degree day information is available.

C301.3(1) has been eliminated and C301.3(2) renumbered to C301.3. Section C301.3 now includes Climate Zone 0 and the revised heating and cooling degree day thresholds for climate zones. Table C301.3 Thermal Climate Zone Definitions has also been revised to include CZ 0 and the adjusted degree days for other climate zones.

PART **3**

Commercial Energy Provisions: Energy Efficiency Requirements

Chapters 4 and 5

- **Chapter 4** Commercial Energy Efficiency
- **Chapter 5** Existing Buildings
  No changes addressed

Chapters 4 and 5 of the *International Energy Conservation Code* (IECC) contain the bulk of the design and construction requirements for commercial buildings and residential buildings greater than three stories in height above grade. Chapter 4 includes provisions for the entire building, including the building envelope and mechanical, service water heating and electrical power and lighting systems. It also includes required additional efficiency packages, an alternative performance compliance path and commissioning requirements. The provisions in Chapter 4 address the portions of the building and building systems that impact energy use and promote the effective use of energy.

Chapter 5 addresses construction of existing buildings – additions, alterations and changes to occupancy – correlating to the requirements of Chapter 4. ■

C401.2
Compliance Paths

C401.3
Envelope Certificate

TABLE C402.1.3
Minimum *R*-values

TABLE C402.1.4
Insulation and Fenestration *U*-Factor Criteria

C402.1.4.1 AND C402.2.1
Roof Insulation and Assembly

TABLE C402.4
U-Factor and SHGC Requirements

C402.5
Air Leakage

C402.5.11
Operable Openings Interlocking

C403
Building Mechanical Systems

C403.4.2.3
Automatic Start and Stop

C403.8.3
Fan Efficiency

C403.8.5
Low-Capacity Ventilation Fans

C405
Electrical Power and Lighting Systems

C405.2.1.2
Occupant Sensor Controls in Warehouse Storage Areas

C405.2.8

Parking Garage Lighting Control

C405.4

Lighting for Plant Growth

C405.11

Automatic Receptacle Control

C405.12

Energy Monitoring

C406

Additional Efficiency Requirements

C407

Total Building Performance

C407.2

Performance-Based Compliance

CHANGE TYPE: Modification

C401.2

Compliance Paths

CHANGE SUMMARY: Compliance path options for the commercial provisions of the IECC are named and the sections required for each option are outlined.

2021 CODE: **C401.2 Application.** Commercial buildings shall comply with Section C401.2.1 or C401.2.2. ~~one of the following:~~

~~The requirements of ANSI/ASHRAE/IESNA 90.1.~~

~~The requirements of Sections C402 through C405 and C408. In addition, commercial buildings shall comply with Section C406 and tenant spaces shall comply with Section C406.1.1.~~

~~The requirements of Sections C402.5, C403.2, C403.3 through C403.3.2, C403.4 through C403.4.2.3, C403.5.5, C403.7, C403.8.1 through C403.8.4, C403.10.1 through C403.10.3, C403.11, C403.12, C404, C405, C407 and C408. The building energy cost shall be equal to or less than 85 percent of the standard reference design building.~~

C401.2.1 International Energy Conservation Code. Commercial buildings shall comply with one of the following:

1. Prescriptive Compliance. The Prescriptive Compliance option requires compliance with Sections C402 through C406 and Section C408. Dwelling units and sleeping units in Group R-2 buildings without systems serving multiple units shall be deemed to be in compliance with this chapter provided that they comply with Section R406.
2. Total Building Performance. The Total Building Performance option requires compliance with Section C407.

Exception: Additions, alterations, repairs and changes of occupancy to existing buildings complying with Chapter 5.

Photo courtesy of Alexander Spatari/Moment/Getty Images

Commercial projects may comply with the IECC or ASHRAE 90.1.

C401.2.2 ASHRAE 90.1. Commercial buildings shall comply with the requirements of ANSI/ASHRAE/IES 90.1.

CHANGE SIGNIFICANCE: Since the 2009 IECC, Section C401.2 has included three options for compliance with the energy code: two IECC compliance options and ASHRAE 90.1. The two IECC options were not named within the code but have commonly been referred to as the prescriptive and performance compliance paths. In the 2012 IECC, the performance compliance option in Section C401.2 required the building energy cost to be less than or equal to 85 percent of the standard reference design. Yet in Section C407, the requirement for the performance path was to have an annual energy cost less than or equal to the annual energy cost of the standard reference design, which led to some confusion among code users.

Section C401.2 has been revised to clarify compliance paths, eliminate conflicting requirements and move existing building requirements to Chapter 5. The IECC compliance options are now named Prescriptive and Total Building Performance, which formalizes the way in which the paths are typically identified and adds clarity for training, education and new code users. Each path is clearly identified as an option, a term that reinforces that it is the designer's choice as to which path is followed. Finally, this change further clarifies that one cannot combine the IECC and ASHRAE compliance paths on a given project.

The list of "mandatory" requirements is eliminated, reducing conflicts with information in Section C407 for Total Building Performance. This corresponds to the creation of a project requirements table in Section C407 and the elimination of the terms *prescriptive* and *mandatory* in provision titles throughout Chapter 4. All requirements associated with the Total Building Performance path are now located in Section C407, increasing usability for those selecting the performance option of the IECC.

C401.3

Envelope Certificate

CHANGE TYPE: Addition

CHANGE SUMMARY: A requirement for the completion of a thermal envelope certificate, which serves as a permanent record of the envelope components installed in the building, is introduced.

2021 CODE: **C401.3 Thermal envelope certificate.** A permanent thermal envelope certificate shall be completed by an approved party. Such certificate shall be posted on a wall in the space where the space conditioning equipment is located, a utility room or other approved location. If located on an electrical panel, the certificate shall not cover or obstruct the visibility of the circuit directory label, service disconnect label or other required labels. A copy of the certificate shall also be included in the construction files for the project. The certificate shall include the following:

1. *R*-values of insulation installed in or on ceilings, roofs, walls, foundations and slabs, basement walls, crawl space walls and floors and ducts outside conditioned spaces.

Energy Efficiency Certificate

Code edition				
Compliance path				
Insulation Rating		***R*-Value**		***R*-Value**
Ceiling/Roof		*R*-		*R*-
Walls	**Frame**	*R*-	**Mass**	*R*-
	Basement	*R*-	**Crawl space**	*R*-
Floors	**Over unconditioned space**	*R*-	**Slab edge**	*R*-
Ducts	**Attic**	*R*-	**Other**	*R*-
Air Leakage Test Results				
Envelope testing	ACH ______ Pa.		**Duct testing**	cfm/100 ft²
Fenestration Rating	**NFRC U-Factor**		**NFRC SHGC**	
Window	*U*-			
Opaque door	*U*-			
Skylight	*U*-			
Weighted average	*U*-			
Designer/builder			**Date**	

This Certificate is to be posted in accordance with Section C401.3 of the International Energy Conservation Code.

Sample thermal envelope certificate.

2. *U*-factors and solar heat gain coefficients (SHGCs) of fenestrations.
3. Results from any building envelope air leakage testing performed on the building.

Where there is more than one value for any component of the building envelope, the certificate shall indicate the area-weighted average value where available. If the area-weighted average is not available, the certificate shall list each value that applies to 10 percent or more of the total component area.

CHANGE SIGNIFICANCE: The addition of C401.3 completes the general requirements of Chapter 4. The new language provides for a permanent certificate for commercial buildings that will record basic information related to the building's thermal envelope. This will complement the information required in Section C408 related to the documentation of mechanical and lighting systems. The certificate requirement is simplified to cover only the basic elements of the thermal envelope, similar to the requirement for residential buildings in Section R401.3, which has been in the IECC since at least the 2006 edition and has been successfully integrated into software programs such as RES*check*.

The information contained in this certificate, while readily available at construction, is often lost as the building ages and ownership is transferred. As future owners or lessors undertake load calculations for HVAC sizing or other measures that require a working knowledge of the building's thermal envelope characteristics, this information will be important. Recording the information in a permanent manner at an approved location in the building, as well as including documentation in the construction files for the project, provides the needed information for future alterations and additions to the building.

This new provision is applicable to both IECC compliance paths.

C402

Building Envelope

Section C402 provides building thermal envelope provisions for commercial buildings, including limitations on air infiltration, or air leakage. The building envelope is important to building energy efficiency. When it is cold outside, heat loss and air leakage through the building envelope add to the heating load. On hot days, solar gains through windows, heat gain through opaque assemblies and the infiltration of hot or humid air contribute to the cooling load. The requirements of Section C402 are intended to reduce these heat gains and losses through the building envelope. Increased envelope efficiency also reduces the initial cost of the mechanical system when the heating and cooling loads of a building are adequately sized, which in turn decreases ongoing heating and cooling related utility costs. Additionally, envelope efficiency contributes to the resiliency of a building; for example, a tight building envelope maintains habitable temperatures for longer periods during power outages.

The commercial envelope provisions have seen few significant changes since the 2012 IECC. Changes to the 2021 IECC are intended to reduce energy use, the single largest operating expense in commercial buildings, representing approximately one-third of typical operating budgets.

These significant changes to the building envelope are addressed in the following sections:

- Increased insulation requirements
- Clarifying language for roof and ceiling assembly insulation
- Reduced fenestration *U*-factors and solar heat gain coefficients (SHGCs), and the elimination of orientation as a factor in these values
- Expanded air leakage requirements
- New operable openings interlocking requirements

Photo courtesy of EasyBuy4u/E+/Getty Images

Efficient building envelopes contribute to the resiliency of the building.

Table C402.1.3

Minimum *R*-Values

CHANGE TYPE: Modification

CHANGE SUMMARY: The minimum *R*-values in Table C402.1.3 are increased to improve the efficiency of the building envelope and to align with ASHRAE 90.1.

2021 CODE: Table C402.1.3

CHANGE SIGNIFICANCE: The *R*-values for many roof, wall and floor components have been increased in Climate Zones 4 through 8 to align with the values in ASHRAE 90.1. These updated *R*-values were deemed cost effective by the ASHRAE envelope committee.

In addition to increased insulation requirements throughout the table, notable changes for Climate Zones 0 through 3 include the addition of R-13 insulation for framed floors in Climate Zones 0 and 1 aligning with the corresponding *U*-factor in Table C402.1.4, and the new R-10 continuous insulation requirement for unheated slabs in Climate Zone 3 Group R buildings. The increased *R*-values reduce tradeoff flexibility for commercial buildings using the Total Building Performance compliance option of Section C407.

Finally, doors have been removed from this table and are listed solely in Table C402.1.4. The proponents of this change based it on cost data gathered from industry partners and thermal performance data from ASHRAE research. The performance of doors is more accurately reflected using *U*-factors, and the industry is moving toward assembly *U*-factors rather than *R*-values for doors.

Table C402.1.3 is shown on the following page. Please refer to the code text for tablenotes and footnotes as these are an integral part of this table. This table is applicable to the Prescriptive Compliance option only.

Photo courtesy of dropStock/iStock / Getty Images Plus/ Getty Images

R-Value is used to measure thermal resistance of insulation.

TABLE C402.1.3 Opaque Thermal Envelope Insulation Component Minimum Requirements, *R*-Value Method

CLIMATE ZONE	0 AND 1		2		3		4 EXCEPT MARINE		5 AND MARINE 4		6		7		8	
	All other	Group R	All other	Group R	All other	Group R	All other	Group R	All other	Group R	All other	Group R	All other	Group R	All other	Group R
Roofs																
Insulation entirely above roof deck	R-20ci	R-25ci	R-25ci	R-25ci	R-25ci	R-25ci	R-30ci	R-30ci	R-30ci	R-30ci	R-30ci	R-30ci	R-35ci	R-35ci	R-35ci	R-35ci
Metal buildings[b]	R-19 + R-11 LS	R-19 + R-11 LS	R-19 + R11 LS	R-19 + R-11 LS	R-19 + R-11 LS	R-19 + R-11 LS	R-19 + R-11 LS	R-19 + R-11 LS	R-19 + R-11 LS	R-19 + R-11 LS	R-25 + R-11 LS	~~R-25 + R-11 LS~~ R-30 + R11 LS	R-30 + R-11 LS	R-30 + R-11 LS	~~R-30 + R-11 LS~~ R-25 + R11+R11 LS	~~R-30 + R-11 LS~~ R-25 + R11+R-11 LS
Attic and other	R-38	R-38	R-38	R-38	R-38	R-38	~~R-38~~ R-49	~~R-38~~ R-49	~~R-38~~ R-49	R-49	R-49	R-49	~~R-49~~ R-60	~~R-49~~ R-60	~~R-49~~ R-60	~~R-49~~ R-60
Walls, above grade																
Mass[f]	R-5.7ci[c]	R-5.7ci[c]	R-5.7ci[c]	R-7.6ci	R-7.6ci	R-9.5ci	R-9.5ci	R-11.4ci	R-11.4ci	R-13.3ci	R-13.3ci	R-15.2ci	R-15.2ci	R-15.2ci	R-25ci	R-25ci
Metal building	R-13+ R-6.5ci	R-13 + R-6.5ci	R13 + R-6.5ci	R-13 + R-13ci	R-13 + R-6.5ci	R-13 + R-13ci	R-13 + R-13ci	~~R-13 + R-13ci~~ R-13+R-14 ci	~~R-13 + R-13ci~~ R-13+R-14 ci	~~R-13 + R-13ci~~ R-13+R-14 ci	~~R-13 + R-13ci~~ R-13+R-14 ci	~~R-13 + R-13ci~~ R-13+R-14 ci	~~R-13 + R-13ci~~ R-13+R-17ci	R-13+ R-19.5ci	~~R-13 + R-13ci~~ R-13+R-19.5ci	R-13+ R-19.5ci
Metal framed	R-13 + R-5ci	R-13 + R-5ci	R-13 + R-5ci	R-13 + R-7.5ci	R-13 + R-7.5ci	R-13 + R-7.5ci	R-13 + R-7.5ci	R-13 + R-7.5ci	~~R-13 + R-7.5ci~~ R-13+R-10ci	~~R-13 + R-7.5ci~~ R-13+R-10ci	~~R-13 + R-7.5ci~~ R-13+R-12.5ci	~~R-13 + R-7.5ci~~ R-13+R-12.5ci	~~R-13 + R-7.5ci~~ R-13+R-12.5ci	R-13 + R-15.6ci	~~R-13 + R-7.5ci~~ R-13+R-18.8ci	~~R-13+ R17.5ci~~ R-13+R-18.8ci
Wood framed and other	R-13 + R-3.8ci or R-20	R-13 + R-3.8ci or R-20	R-13 + R-3.8ci or R-20	R-13 + R-3.8ci or R-20	R-13 + R-3.8ci or R-20	R-13 + R-3.8ci or R-20	R-13 + R-3.8ci or R-20	R-13 + R-3.8ci or R-20	~~R-13 + R-3.8ci or R-20~~ R-13+R-7.5ci or R-20+R-3.8ci	R-13 + R-7.5ci or R-20 + R-3.8ci	R-13 + R-7.5ci or R-20 + R-3.8ci	R-13 + R-7.5ci or R-20 + R-3.8ci	R-13 + R-7.5ci or R-20 + R-3.8ci	R-13 + R-7.5ci or R-20 + R-3.8ci	~~R13 + R-15.6ci or R-20 + R-10ci~~ R-13+R18.8ci	~~R13 + R-15.6ci or R-20 + R-10ci~~ R-13+R18.8ci
Walls, below grade																
Below-grade wall[d]	NR	NR	NR	NR	NR	NR	R-7.5ci	~~R-7.5ci~~ R-10ci	R-7.5ci	~~R-7.5ci~~ R-10ci	~~R-7.5ci~~ R-10ci	~~R-7.5ci~~ R-15ci	~~R-10ci~~ R-15ci	~~R-10ci~~ R-15ci	~~R-10ci~~ R-15ci	~~R-12.5ci~~ R-15ci
Floors																
Mass[e]	NR	NR	R-6.3ci	R-8.3ci	R-10ci	R-10ci	~~R-10ci~~ R-14.6ci	~~R-10.4ci~~ R-16.7ci	~~R-10ci~~ R-14.6ci	~~R-12.5ci~~ R-16.7ci	~~R-12.5ci~~ R-16.7ci	~~R-12.5ci~~ R-16.7ci	~~R-15ci~~ R-20.9ci	~~R-16.7ci~~ R20.9ci	~~R-15ci~~ R-23ci	~~R-16.7ci~~ R-23ci
Joist/framing	~~NR~~ R-13	~~NR~~ R-13	R-30	R-30	R-30	R-30	R-30	R-30	R-30	R-30	~~R-30~~ R-38	~~R-30[f]~~ R-38	~~R-30[f]~~ R-38	~~R-30[f]~~ R-38	~~R-30[f]~~ R-38	~~R-30[f]~~ R-38
Slab-on-grade floors																
Unheated slabs[g]	NR	NR	NR	NR	NR	~~NR~~ R-10 for 24" below	~~R-10 for 24" below~~ R-15 for 24" below	~~R-10 for 24" below~~ R-15 for 24" below	~~R-10 for 24" below~~ R-15 for 24" below	~~R-10 for 24" below~~ R-20 for 24" below	~~R-10 for 24" below~~ R-20 for 24" below	~~R-15 for 24" below~~ R-20 for 48" below	~~R-15 for 24" below~~ R-20 for 24" below	~~R-15 for 24" below~~ R-20 for 48" below	~~R-15 for 24" below~~ R-20 for 48" below	~~R-20 for 24" below~~ R-25 for 48" below
Heated slabs[b]	R-7.5 for 12" below + R-5 full slab	R-7.5 for 12" below + R-5 full slab	R-7.5 for 12" below + R-5 full slab	R-7.5 for 12" below + R-5 full slab	R-10 for 24" below + R-5 full slab	R-10 for 24" below + R-5 full slab	R-15 for 24" below + R-5 full slab	R-15 for 24" below + R-5 full slab	R-15 for 36" below + R-5 full slab	R-15 for 36" below + R-5 full slab	R-15 for 36" below + R-5 full slab	R-20 for 48" below + R-5 full slab	R-20 for 48" below + R-5 full slab	R-20 for 48" below + R-5 full slab	R-20 for 48" below + R-5 full slab	R-20 for 48" below + R-5 full slab
~~**Opaque doors**~~																
~~Nonswinging~~	~~R-4.75~~	~~R-4.75~~	~~R-4.75~~	~~R-4.75~~	~~R-4.75~~	~~R-4.75~~	~~R-4.75~~	~~R-4.75~~	~~R-4.75~~	~~R-4.75~~	~~R-4.75~~	~~R-4.75~~	~~R-4.75~~	~~R-4.75~~	~~R-4.75~~	~~R-4.75~~

Table C402.1.4

Insulation and Fenestration *U*-Factor Criteria

CHANGE TYPE: Modification

CHANGE SUMMARY: The *U*-factor requirements in Table C402.1.4 are decreased to improve the efficiency of the building envelope and to align with ASHRAE 90.1.

2021 CODE: Table C402.1.4

CHANGE SIGNIFICANCE: The assembly *U*-factors for many roof, wall and floor assemblies have been decreased in Climate Zones 4 through 8 to align with the values in ASHRAE 90.1. These assemblies were deemed cost effective by the ASHRAE envelope committee.

Previously, the *F*-factors for heated slabs in Table C402.1.4 were incorrectly shown as two separate factors for the two insulation components; this has been corrected in the 2021 IECC and now only one *F*-factor is shown. *F*-factors are like *U*-factors in that they represent the net effect of all insulation materials applied to an assembly, and there should only be one *F*-factor representing the *R*-value of the insulation materials used on the slab construction.

As with the changes to Table C402.1.3, the change to maximum *U*-factors reduces tradeoff flexibility for commercial buildings using the Total Building Performance compliance option of Section C407.

Finally, all door *U*-factor values are consolidated into this table. The proponents of this change based it on cost data gathered from industry partners and thermal performance data from ASHRAE research. The performance of doors is more accurately reflected using *U*-factors, and the industry is moving toward assembly *U*-factors rather than *R*-values for doors.

This table is applicable to the Prescriptive Compliance option only.

Photo courtesy of smodj/iStock / Getty Images Plus/ Getty Images

U-factor is used to measure thermal transmittance of fenestration and envelope assemblies.

TABLE C402.1.4 Opaque Thermal Envelope Assembly Maximum Requirements, *U*-Factor Method[a,b]

CLIMATE ZONE	0 AND 1		2		3		4 EXCEPT MARINE		5 AND MARINE 4		6		7		8	
	All other	Group R	All other	Group R	All other	Group R	All other	Group R	All other	Group R	All other	Group R	All other	Group R	All other	Group R
Roofs																
Insulation entirely above roof deck	U-0.048	U-0.039	U-0.039	U-0.039	U-0.039	U-0.039	U-0.032	U-0.032	U-0.032	U-0.032	U-0.032	U-0.032	U-0.028	U-0.028	U-0.028	U-0.028
Metal buildings	~~U-0.044~~ U-0.035	U-0.035	U-0.035	U-0.035	U-0.035	U-0.035	U-0.035	U-0.035	U-0.035	U-0.035	U-0.031	~~U-0.031~~ U-0.029	U-0.029	U-0.029	~~U-0.029~~ U-0.026	~~U-0.029~~ U-0.026
Attic and other	U-0.027	U-0.027	U-0.027	U-0.027	U-0.027	U-0.027	~~U-0.027~~ U-0.021	~~U-0.027~~ U-0.021	~~U-0.027~~ U-0.021	U-0.021	U-0.021	U-0.021	~~U-0.021~~ U-0.017	~~U-0.021~~ U-0.017	~~U-0.021~~ U-0.017	~~U-0.021~~ U-0.017
Walls, above grade																
Mass[g]	U-0.151	U-0.151	U-0.151	U-0.123	U-0.123	U-0.104	U-0.104	U-0.090	U-0.090	U-0.080	U-0.080	U-0.071	U-0.071	U-0.071	~~U-0.061~~ U-0.037	~~U-0.061~~ U-0.037
Metal building	U-0.079	U-0.079	U-0.079	U-0.079	U-0.079	U-0.052	U-0.052	~~U-0.052~~ U-0.050	~~U-0.052~~ U-0.050	~~U-0.052~~ U-0.050	~~U-0.052~~ U-0.050	~~U-0.052~~ U-0.050	~~U-0.052~~ U-0.044	U-0.039	~~U-0.052~~ U-0.039	U-0.039
Metal framed	U-0.077	U-0.077	U-0.077	U-0.064	U-0.064	U-0.064	U-0.064	U-0.064	~~U-0.064~~ U-0.055	~~U-0.064~~ U-0.055	~~U-0.064~~ U-0.049	~~U-0.064~~ U-0.049	~~U-0.064~~ U-0.049	~~U-0.052~~ U-0.042	~~U-0.064~~ U-0.037	~~U-0.045~~ U-0.037
Wood framed and other[c]	U-0.064	U-0.064	U-0.064	U-0.064	U-0.064	U-0.064	U-0.064	U-0.064	~~U-0.064~~ U-0.051	~~U-0.064~~ U-0.051	U-0.051	U-0.051	U-0.051	U-0.051	~~U-0.036~~ U-0.032	~~U-0.036~~ U-0.032
Walls, below grade																
Below-grade wall[c]	C-1.140[e]	C-1.140[e]	C-1.140[e]	C-1.140[e]	C-1.140[e]	C-1.140[e]	C-0.119	~~C-0.119~~ C-0.092	C-0.119	~~C-0.119~~ C-0.092	~~C-0.119~~ C-0.092	~~C-0.119~~ C-0.063	~~C-0.092~~ C-0.063	~~C-0.092~~ C-0.063	~~C-0.092~~ C-0.063	~~C-0.092~~ C-0.063
Floors																
Mass[d]	U-0.322[e]	U-0.322[e]	U-0.107	U-0.087	~~U-0.076~~ U-0.074	~~U-0.076~~ U-0.074	~~U-0.076~~ U-0.057	~~U-0.074~~ U-0.051	~~U-0.074~~ U-0.057	~~U-0.064~~ U-0.051	~~U-0.064~~ U-0.051	~~U-0.064~~ U-0.051	~~U-0.055~~ U-0.042	~~U-0.051~~ U-0.042	~~U-0.055~~ U-0.038	~~U-0.051~~ U-0.038
Joist/framing	U-0.066[e]	U-0.066[e]	U-0.033	U-0.033	U-0.033	U-0.033	U-0.033	U-0.033	U-0.033	U-0.033	~~U-0.033~~ U-0.027	~~U-0.033~~ U-0.027	~~U-0.033~~ U-0.027	~~U-0.033~~ U-0.027	~~U-0.033~~ U-0.027	~~U-0.033~~ U-0.027

(Continued)

TABLE C402.1.4 (*Continued*)

CLIMATE ZONE	0-1		2		3		4 EXCEPT MARINE		5 AND MARINE 4		6		7		8	
	All other	Group R	All other	Group R	All other	Group R	All other	Group R	All other	Group R	All other	Group R	All other	Group R	All other	Group R
Slab-on-grade floors																
Unheated slabs	F-0.73[e]	F-0.73[e]	F-0.73[e]	F-0.73[e]	F-0.73[e]	~~F-0.73~~ F-0.54	~~F-0.54~~ F-0.52	~~F-0.54~~ F-0.52	~~F-0.54~~ F-0.52	~~F-0.54~~ F-0.51	~~F-0.54~~ F-0.51	~~F-0.52~~ F-0.434	~~F-0.40~~ F-0.51	~~F-0.40~~ F-0.434	~~F-0.40~~ F-0.434	~~F-0.40~~ F-0.424
Heated slabs[f]	~~F-1.02~~ ~~0.74~~ F-0.69	~~F-1.02~~ ~~0.74~~ F-0.69	~~F-1.02~~ ~~0.74~~ F-0.69	~~F-1.02~~ ~~0.74~~ F-0.69	~~F-0.90~~ ~~0.74~~ F-0.66	~~F-0.90~~ ~~0.74~~ F-0.66	~~F-0.86~~ ~~0.64~~ F-0.62	~~F-0.86~~ ~~0.64~~ F-0.62	~~F-0.79~~ ~~0.64~~ F-0.62	~~F-0.79~~ ~~0.64~~ F-0.62	~~F-0.79~~ ~~0.55~~ F-0.62	~~F-0.69~~ ~~0.55~~ F-0.602	~~F-0.69~~ ~~0.55~~ F-0.602	~~F-0.69~~ ~~0.55~~ F-0.602	~~F-0.69~~ ~~0.55~~ F-0.602	~~F-0.69~~ ~~0.55~~ F-0.602
Opaque doors																
Nonswinging door	U-0.31	U-0.31	U-0.31	U-0.31	U-0.31	U-0.31	U-0.31	U-0.31	U-0.31	U-0.31	U-0.31	U-0.31	U-0.31	U-0.31	U-0.31	U-0.31
Swinging door[g]	~~U-0.61~~ U-0.37	~~U-0.61~~ U-0.37	~~U-0.61~~ U-0.37	~~U-0.61~~ U-0.37	~~U-0.61~~ U-0.37	~~U-0.61~~ U-0.37	~~U-0.61~~ U-0.37	~~U-0.61~~ U-0.37	U-0.37	U-0.37	U-0.37	U-0.37	U-0.37	U-0.37	U-0.37	U-0.37
Garage door <14% glazing[h]	U-0.31	U-0.31	U-0.31	U-0.31	U-0.31	U-0.31	U-0.31	U-0.31	U-0.31	U-0.31	U-0.31	U-0.31	U-0.31	U-0.31	U-0.31	U-0.31

For SI: 1 pound per square foot = 4.88 kg/m^2, 1 pound per cubic foot = 16 kg/m^3.

ci = Continuous Insulation, NR = No Requirement, LS = Liner System.

a. Where assembly *U*-factors, *C*-factors and *F*-factors are established in ANSI/ASHRAE/IES 90.1 Appendix A, such opaque assemblies shall be a compliance alternative where those values meet the criteria of this table, and provided that the construction, excluding the cladding system on walls, complies with the appropriate construction details from ANSI/ASHRAE/IES 90.1 Appendix A.

b. Where *U*-factors have been established by testing in accordance with ASTM C1363, such opaque assemblies shall be a compliance alternative where those values meet the criteria of this table. The *R*-value of continuous insulation shall be permitted to be added to or subtracted from the original tested design.

c. Where heated slabs are below grade, below-grade walls shall comply with the *U*-factor requirements for above-grade mass walls.

d. "Mass floors" shall be in accordance with Section C402.2.3.

e. These *C*-, *F*- and *U*-factors are based on assemblies that are not required to contain insulation.

~~f. The first value is for perimeter insulation and the second value is for full, **under-slab** insulation.~~

f. "Mass walls" shall be in accordance with Section C402.2.2.

g. Swinging door *U*-factors shall be determined in accordance with NFRC-100.

h. Garage doors having a single row of fenestration shall have an assembly *U*-factor less than or equal to 0.44 in Climate Zones 0 through 6 and less than or equal to 0.36 in Climate Zones 7 and 8, provided that the fenestration area is not less than 14 percent and not more than 25 percent of the total door area.

C402.1.4.1 and C402.2.1

Roof Assembly and Insulation

CHANGE TYPE: Modification

CHANGE SUMMARY: Roof insulation requirements are clarified and separated for compliance with either the *U*-factor or *R*-value method in Sections C402.1.4.1 and C402.2.1.

2021 CODE: **C402.1.4.1 Roof/ceiling assembly.** The maximum roof/ceiling assembly *U*-factor shall not exceed that specified in Table C402.1.4 based on construction materials used in the roof/ceiling assembly.

C402.1.4.1.1 Tapered, above-deck insulation based on thickness. Where used as a component of a maximum roof/ceiling assembly *U*-factor calculation, the sloped roof insulation *R*-value contribution to that calculation shall use the average thickness in inches (mm) along with the material *R*-value-per-inch (per-mm) solely for *U*-factor compliance as prescribed in Section C402.1.4.

C402.1.4.1.2 Suspended ceilings. Insulation installed on suspended ceilings having removable ceiling tiles shall not be considered part of the assembly *U*-factor of the roof/ceiling construction.

C402.1.4.1.3 Joints staggered. Continuous insulation board shall be installed in not less than two layers, and the edge joints between each layer of insulation shall be staggered, except where insulation tapers to the roof deck at a gutter edge, roof drain or scupper.

C402.2.1 Roof assembly. The minimum thermal resistance (*R*-value) of the insulating material installed either between the roof framing or continuously on the roof assembly shall be as specified in Table C402.1.3, based on construction materials used in the roof assembly. ~~Insulation installed on a suspended ceiling having removable ceiling tiles shall not be considered~~

Photo courtesy of tzahiV/iStock / Getty Images Plus/ Getty Images

Roof insulation may comply with the *R*-value or *U*-factor method.

~~as part of the minimum thermal resistance of the roof insulation. Continuous insulation board shall be installed in not less than 2 layers and the edge joints between each layer of insulation shall be staggered.~~

~~**Exceptions:**~~

1. ~~Continuously insulated roof assemblies where the thickness of insulation varies 1 inch (25 mm) or less and where the area-weighted *U*-factor is equivalent to the same assembly with the *R*-value specified in Table C402.1.3.~~
2. ~~Where tapered insulation is used with insulation entirely above deck, the *R*-value where the insulation thickness varies 1 inch (25 mm) or less from the minimum thickness of tapered insulation shall comply with the *R*-value specified in Table C402.1.3.~~
3. ~~Two layers of insulation are not required where insulation tapers to the roof deck, such as at roof drains.~~

C402.2.1.1 Tapered, above-deck insulation based on thickness. Where used as a component of a roof/ceiling assembly *R*-value calculation, the sloped roof insulation *R*-value contribution to that calculation shall use the average thickness in inches (mm) along with the material *R*-value-per-inch (per-mm) solely for *R*-value compliance as prescribed in Section 402.1.3.

C402.2.1.2 Minimum thickness, lowest point. The minimum thickness of above-deck roof insulation at its lowest point, gutter edge, roof drain or scupper, shall be no less than 1 inch (25 mm).

C402.2.1.3 Suspended ceilings. Insulation installed on suspended ceilings having removable ceiling tiles shall not be considered part of the minimum thermal resistance (*R*-value) of roof insulation in roof/ceiling construction.

C402.2.1.4 Joints staggered. Continuous insulation board shall be installed in not less than two layers and the edge joints between each layer of insulation shall be staggered, except where insulation tapers to the roof deck at a gutter edge, roof drain or scupper.

CHANGE SIGNIFICANCE: The area-weighted average *U*-factor option in lieu of *R*-value compliance for roofs was introduced in the 2006 IECC, and the next code cycle brought forth the equivalent *U*-factor option most users are now accustomed to. However, changes to the 2018 IECC language resulted in confusion as to the application of weighted averages, minimum insulation thickness and exceptions associated with area-weighted *U*-factors under *R*-value requirements.

The 2021 IECC provisions provide clarity on roof/ceiling assembly requirements as they relate to *U*-factors and *R*-values. The new Section C402.1.4.1 includes provisions for roof/ceiling assemblies using a *U*-factor approach, while the new subsections of Section C402.2.1 include provisions for roof/ceiling assemblies using the *R*-value approach. The changes support improved application and enforcement of roof/ceiling assembly insulation requirements.

These sections are only applicable to the Prescriptive Compliance option.

Table C402.4

U-Factor and SHGC Requirements

CHANGE TYPE: Modification

CHANGE SUMMARY: Changes to the *U*-factor and SHGC requirements in Table C402.4 increase the energy efficiency of windows, doors and skylights, and the table is revised to classify SHGC on the operable/fixed properties of the fenestration rather than orientation.

2021 CODE: Table C402.4

CHANGE SIGNIFICANCE: Commercial fenestration *U*-factors remained unchanged between the 2012 and 2018 versions of the IECC. SHGC values decreased slightly between versions, and the 2015 IECC saw the return of orientation as a component of SHGC. Within the 2021 IECC is a widespread reduction in fenestration *U*-factors and SHGCs, and orientation is once again eliminated as a factor for SHGC in the Prescriptive Compliance path.

The changes were analyzed by Pacific Northwest National Laboratory (PNNL) to determine what performance levels showed positive life cycle energy savings. Depending on the climate zone, *U*-factors are reduced by 0 to 14 percent for fixed and operable vertical fenestration, 7 to 25 percent for entrance doors and 0 to 18 percent for skylights; SHGC requirements for vertical fenestration and skylights are similarly reduced by 0 to 14 percent.

The elimination of orientation from the table simplifies compliance with and enforcement of the prescriptive fenestration SHGC requirements in the IECC for architects, designers and code officials by adopting a "fixed" and "operable" approach to setting SHGC requirements, rather than determining the orientation of each fenestration product in the building.

The projection factor adjustments are retained but modified to correspond with the new "fixed" and "operable" distinction. Analysis by PNNL indicated these changes result in energy savings and peak demand savings in every climate zone, and, in many cases, reduce the size of cooling equipment. The revision also brings greater consistency between IECC and ASHRAE 90.1 SHGC requirements.

Table C402.4 only applies to the Prescriptive Compliance option. To the extent that design professionals want to factor orientation into a more sophisticated design strategy, the Section C407 Total Building Performance compliance path can be used.

Photo courtesy of uschools/E+/Getty Images Plus

U-Factor and SHGC requirements indicate heat transmittance of fenestration.

TABLE C402.4 Building Envelope Fenestration Maximum *U*-Factor and SHGC Requirements

CLIMATE ZONE	0 AND 1		2		3		4 EXCEPT MARINE		5 AND MARINE 4		6		7		8	
Vertical fenestration																
***U*-factor**																
Fixed fenestration	0.50		~~0.50~~ 0.45		~~0.46~~ 0.42		~~0.38~~ 0.36		~~0.38~~ 0.36		~~0.36~~ 0.34		0.29		~~0.29~~ 0.26	
Operable fenestration	~~0.65~~ 0.62		~~0.65~~ 0.60		~~0.60~~ 0.54		0.45		0.45		~~0.43~~ 0.42		~~0.37~~ 0.36		~~0.37~~ 0.32	
Entrance doors	~~1.10~~ 0.83		~~0.83~~ 0.77		~~0.77~~ 0.68		~~0.77~~ 0.63		~~0.77~~ 0.63		~~0.77~~ 0.63		~~0.77~~ 0.63		~~0.77~~ 0.63	
SHGC																
~~Orientation~~[a]	~~SEW~~ Fixed	~~N~~ Operable	~~SEW~~ Fixed	~~N~~ Operable	~~SEW~~ Fixed	~~N~~ Operable	~~SEW~~ Fixed	~~N~~ Operable	~~SEW~~ Fixed	~~N~~ Operable	~~SEW~~ Fixed	~~N~~ Operable	~~SEW~~ Fixed	~~N~~ Operable	~~SEW~~ Fixed	~~N~~ Operable
PF < 0.2	~~0.25~~ 0.23	~~0.33~~ 0.21	0.25	~~0.33~~ 0.23	0.25	~~0.33~~ 0.23	0.36	~~0.48~~ 0.33	0.38	~~0.51~~ 0.33	~~0.40~~ 0.38	~~0.53~~ 0.34	~~0.45~~ 0.40	~~NR~~ 0.36	~~0.45~~ 0.40	~~N~~ 0.36
0.2 ≤ PF < 0.5	~~0.30~~ 0.28	~~0.37~~ 0.25	0.30	~~0.37~~ 0.28	0.30	~~0.37~~ 0.28	0.43	~~0.53~~ 0.40	0.46	~~0.56~~ 0.40	~~0.48~~ 0.46	~~0.58~~ 0.41	~~NR~~ 0.48	~~NR~~ 0.43	~~NR~~ 0.48	~~NR~~ 0.43
PF ≥ 0.5	~~0.40~~ 0.37	~~0.40~~ 0.34	0.40	~~0.40~~ 0.37	0.40	~~0.40~~ 0.37	0.58	~~0.58~~ 0.53	0.61	~~0.61~~ 0.53	~~0.64~~ 0.61	~~0.64~~ 0.54	~~NR~~ 0.64	~~NR~~ 0.58	~~NR~~ 0.64	~~NR~~ 0.58
Skylights																
U-factor	~~0.75~~ 0.70		0.65		0.55		0.50		0.50		0.50		~~0.50~~ 0.44		~~0.50~~ 0.41	
SHGC	~~0.35~~ 0.30		~~0.35~~ 0.30		~~0.35~~ 0.30		0.40		0.40		0.40		NR		NR	

NR = No Requirement, PF = Projection Factor.

~~a. "N" indicates vertical fenestration oriented within 45 degrees of true north. "SEW" indicates orientations other than "N." For buildings in the southern hemisphere, reverse south and north. Buildings located at less than 23.5 degrees latitude shall use SEW for all orientations.~~

C402.5

Air Leakage

CHANGE TYPE: Modification

CHANGE SUMMARY: In Section C402.5, air leakage requirements are expanded to include requirements for residential and nonresidential air leakage testing and for building envelope performance verification for buildings not tested.

2021 CODE: C402.5 Air leakage—thermal envelope ~~(Mandatory).~~ The building thermal envelope ~~of buildings~~ shall comply with Sections C402.5.1 through C402.5.11.1, or the building thermal envelope shall be tested in accordance with Section C402.5.2 or C402.5.3. ~~ASTM E 779 at a pressure differential of 0.3 inch water gauge (75 Pa) or an equivalent method approved by the code official and deemed to comply with the provisions of this section when the tested air leakage rate of the building thermal envelope is not greater than 0.40 cfm/ft² (2.0 L/s · m²).~~ Where compliance is based on such testing, the building shall also comply with Sections C402.5.7, C402.5.8 and C402.5.9.

C402.5.1 Air barriers. A continuous air barrier shall be provided throughout the building thermal envelope. The continuous air barriers shall ~~be permitted to~~ be located on the inside or outside of the building thermal envelope, located within the assemblies composing the building thermal envelope, or any combination thereof. The air barrier shall comply with Sections C402.5.1.1 and C402.5.1.2.

> **Exception:** Air barriers are not required in buildings located in Climate Zone 2B.

C402.5.1.1 Air barrier construction. The continuous air barrier shall be constructed to comply with the following:

1. The air barrier shall be continuous for all assemblies that are the thermal envelope of the building and across the joints and assemblies.

Expanded air leakage provisions include testing requirements.

2. Air barrier joints and seams shall be sealed, including sealing transitions in places and changes in materials. The joints and seals shall be securely installed in or on the joint for its entire length so as not to dislodge, loosen or otherwise impair its ability to resist positive and negative pressure from wind, stack effect and mechanical ventilation.
3. Penetrations of the air barrier shall be caulked, gasketed or otherwise sealed in a manner compatible with the construction materials and location. Sealing shall allow for expansion, contraction and mechanical vibration. Joints and seams associated with penetrations shall be sealed in the same manner or taped. Sealing materials shall be securely installed around the penetration so as not to dislodge, loosen or otherwise impair the penetrations' ability to resist positive and negative pressure from wind, stack effect and mechanical ventilation. Sealing of concealed fire sprinklers, where required, shall be in a manner that is recommended by the manufacturer. Caulking or other adhesive sealants shall not be used to fill voids between fire sprinkler cover plates and walls or ceilings.
4. Recessed lighting fixtures shall comply with Section C402.5.10. Where similar objects are installed that penetrate the air barrier, provisions shall be made to maintain the integrity of the air barrier.

C402.5.1.2 Air barrier compliance ~~options~~. A continuous air barrier for the opaque building envelope shall comply with ~~Section C402.5.1.2.1 or C402.5.1.2.2.~~ the following:

1. Buildings or portions of buildings, including Group R and I occupancies, shall meet the provisions of Section C402.5.2.

   Exception: Buildings in Climate Zones 2B, 3C and 5C.

2. Buildings or portions of buildings other than Group R and I occupancies shall meet the provisions of Section C402.5.3.

   Exceptions:

   1. Buildings in Climate Zones 2B, 3B, 3C and 5C.
   2. Buildings larger than 5,000 square feet (464.5 m^2) floor area in Climate Zones 0B, 1, 2A, 4B and 4C.
   3. Buildings between 5,000 square feet (464.5 m^2) and 50,000 square feet (4645 m^2) floor area in Climate Zones 0A, 3A and 5B.

3. Buildings or portions of buildings that do not complete air barrier testing shall meet the provisions of Section C402.5.1.3 or C402.5.1.4 in addition to Section C402.5.1.5.

~~C402.5.1.2.1~~ C402.5.1.3 Materials. Materials with an air permeability not greater than 0.004 cfm/ft^2 (0.02 L/s × m^2) under a pressure differential of 0.3 inch water gauge (75 Pa) when tested in accordance with ASTM E2178 shall comply with this section. Materials in Items 1 through 16 shall be deemed to comply with this section, provided that joints are sealed, and materials are installed as air barriers in accordance with the manufacturer's instructions.

1. Plywood with a thickness of not less than 3/8 inch (10 mm).
2. Oriented strand board having a thickness of not less than 3/8 inch (10 mm).

3. Extruded polystyrene insulation board having a thickness of not less than 1/2 inch (12.7 mm).
4. Foil-back polyisocyanurate insulation board having a thickness of not less than 1/2 inch (12.7 mm).
5. Closed-cell spray foam having a minimum density of 1.5 pcf (2.4 kg/m^3) and having a thickness of not less than 11/2 inches (38 mm).
6. Open-cell spray foam with a density between 0.4 and 1.5 pcf (0.6 and 2.4 kg/m^3) and having a thickness of not less than 4.5 inches (113 mm).
7. Exterior or interior gypsum board having a thickness of not less than 1/2 inch (12.7 mm).
8. Cement board having a thickness of not less than 1/2 inch (12.7 mm).
9. Built-up roofing membrane.
10. Modified bituminous roof membrane.
11. ~~Fully adhered single-ply~~ Single-ply roof membrane.
12. A Portland cement/sand parge, or gypsum plaster having a thickness of not less than 5/8 inch (15.9 mm).
13. Cast-in-place and precast concrete.
14. Fully grouted concrete block masonry.
15. Sheet steel or aluminum.
16. Solid or hollow masonry constructed of clay or shale masonry units.

~~C402.5.1.2.2~~ C402.5.1.4 Assemblies. Assemblies of materials and components with an average air leakage not greater than 0.04 cfm/ft^2 (0.2 L/s × m^2) under a pressure differential of 0.3 inch of water gauge (w.g.) (75 Pa) when tested in accordance with ASTM E2357, ASTM E1677, ASTM D8052 or ASTM E283 shall comply with this section. Assemblies listed in Items 1 through 3 shall be deemed to comply, provided that joints are sealed and the requirements of Section C402.5.1.1 are met.

1. Concrete masonry walls coated with either one application of block filler or two applications of a paint or sealer coating.
2. Masonry walls constructed of clay or shale masonry units with a nominal width of 4 inches (102 mm) or more.
3. A Portland cement/sand parge, stucco or plaster not less than 1/2 inch (12.7 mm) in thickness.

C402.5.1.5 Building envelope performance verification. The installation of the continuous air barrier shall be verified by the code official, a registered design professional or approved agency in accordance with the following:

- A review of the construction documents and other supporting data shall be conducted to assess compliance with the requirements in Sections C402.5.1.
- Inspection of continuous air barrier components and assemblies shall be conducted during construction while the air barrier is still accessible for inspection and repair to verify compliance with the requirements of Sections C402.5.1.3 and C402.5.1.4.
- A final commissioning report shall be provided for inspections completed by the registered design professional or approved agency. The commissioning report shall be provided to the building owner

or owner's authorized agent and the code official. The report shall identify deficiencies found during the review of the construction documents and inspection and details of corrective measures taken.

C402.5.2 Dwelling and sleeping unit enclosure testing. The building thermal envelope shall be tested in accordance with ASTM E 779, ANSI/RESNET/ICC 380, ASTM E1827 or an equivalent method approved by the code official. The measured air leakage shall not exceed 0.30 cfm/ft^2 (1.5 L/s $\times$ m^2) of the testing unit enclosure area at a pressure differential of 0.2 inch water gauge (50 Pa). Where multiple dwelling units or sleeping units or other occupiable conditioned spaces are contained within one building thermal envelope, each unit shall be considered an individual testing unit, and the building air leakage shall be the weighted average of all testing unit results, weighted by each testing unit's enclosure area. Units shall be tested separately with an unguarded blower door test as follows:

1. Where buildings have fewer than eight testing units, each testing unit shall be tested.
2. For buildings with eight or more testing units, the greater of seven units or 20 percent of the testing units in the building shall be tested, including a top floor unit, a ground floor unit and a unit with the largest testing unit enclosure area. For each tested unit that exceeds the maximum air leakage rate, an additional two units shall be tested, including a mixture of testing unit types and locations.

C402.5.3 Building thermal envelope testing The building thermal envelope shall be tested in accordance with ASTM E779, ANSI/RESNET/ICC 380, ASTM E3158 or ASTM E1827 or an equivalent method approved by the code official. The measured air leakage shall not exceed 0.40 cfm/ft^2 (2.0 L/s $\times$ m^2) of the building thermal envelope area at a pressure differential of 0.3 inch water gauge (75 Pa). Alternatively, portions of the building shall be tested and the measured air leakages shall be area weighted by the surface areas of the building envelope in each portion. The weighted average test results shall not exceed the whole building leakage limit. In the alternative approach, the following portions of the building shall be tested:

1. The entire envelope area of all stories that have any spaces directly under a roof.
2. The entire envelope area of all stories that have a building entrance, exposed floor, or loading dock, or are below grade.
3. Representative above-grade sections of the building totaling at least 25 percent of the wall area enclosing the remaining conditioned space.

Exception: Where the measured air leakage rate exceeds 0.40 cfm/ft^2 (2.0 L/s $\times$ m^2) but does not exceed 0.60 cfm/ft^2 (3.0 L/s $\times$ m^2), a diagnostic evaluation using smoke tracer or infrared imaging shall be conducted while the building is pressurized along with a visual inspection of the air barrier. Any leaks noted shall be sealed where such sealing can be made without destruction of existing building components. An additional report identifying the corrective actions taken to seal leaks shall be submitted to the code official and the building owner, and shall be deemed to comply with the requirements of this section.

CHANGE SIGNIFICANCE: Air leakage testing was introduced as a compliance alternative to meeting the material or assembly selection and installation provisions of the 2012 IECC, and the requirements remained largely unchanged until the expansion of requirements in the 2021 IECC. Three significant changes related to air leakage are made to the 2021 IECC:

1. New requirements for dwelling unit air leakage testing for Group R and I occupancies
2. Revised and required air leakage requirements for occupancies other than Group R and I
3. New performance verification requirements

Air leakage can be a significant source of energy waste in buildings, contributing to higher heating and cooling costs for building owners and occupants and impacting comfort and durability. Airtightness testing often results in increased attention to air barrier sealing and significantly reduced air leakage.

Adequate control over air leakage can provide many benefits, including reduced HVAC equipment sizing, better building pressurization and energy savings due to reduced heating and cooling of infiltrated outside air.

In moist climates, ensuring lower air leakage through whole building testing can also result in better humidity control and reduced risk of durability issues. Pacific Northwest National Laboratory's technical brief, titled, "Envelope Air Tightness for Commercial Buildings" notes:

> *While it is important that the materials and assemblies have limited leakage, that alone does not guarantee a low leakage building. Recent research[1] (Wiss 2014) shows that 40% of buildings constructed without an envelope consultant have air leakage exceeding the currently optional test standard requirements, while buildings with envelope consultants all had leakage below 0.25 cfm/ft^2. Testing is the most reliable means of ensuring that the intent of this code section—limiting unintended energy waste in buildings due to air infiltration—will be achieved.*

The updated air leakage testing requirements of Section C402.5 take into consideration occupancy type, climate zone and building size. Commercial buildings under 5,000 square feet can be tested using residential methods, technicians and equipment with the maximum leakage rate set at 0.30 cfm/ft^2 (1.5 L/s $\times$ m^2) at 0.2 in. w.g. (50 Pa). This testing pressure differential is common for residential testing and is equivalent to a leakage rate of 0.40 cfm/ft^2 (1.5 L/s $\times$ m^2) at 0.3 in. w.g. (75 Pa). Implementing the residential procedure can significantly reduce testing costs for these smaller buildings. Buildings that are not tested must meet the materials or assemblies requirements and the air barrier must be visually inspected. A final commissioning report is required for such inspections.

There are instances when the building is tested and it exceeds the maximum leakage rate. Section C402.5.3 provides reasonable options for mitigating air leakage when the rate is greater than allowed in the code but does not exceed 0.60 cfm/ft^2, including the use of a smoke tracer or infrared imaging along with a visual inspection. Leaks must be sealed where it is possible to do so without destroying building components. Documentation showing all leaks that were found and mitigating measures must be submitted to the code official and building owner.

These provisions are applicable to both IECC compliance paths.

1 Wiss J. 2014. *ASHRAE 1478-RP Measuring Airtightness of Mid- and High-Rise Non-Residential Buildings.* Elstner Associates, Inc. for ASHRAE. https://www.ashrae.org/resources--publications/periodicals/enewsletters/esociety/2014-12-10-articles/completed-research-december-2014.

C402.5.11

Operable Openings Interlocking

CHANGE TYPE: Addition

CHANGE SUMMARY: Large, operable openings such as roll-up doors and windows must now be interlocked with the heating and cooling system.

2021 CODE: **C402.5.11 Operable openings interlocking.** Where occupancies utilize operable openings to the outdoors that are larger than 40 square feet (3.7 m^2) in area, such openings shall be interlocked with the heating and cooling system so as to raise the cooling setpoint to 90°F (32°C) and lower the heating setpoint to 55°F (13°C) whenever the operable opening is open. The change in heating and cooling setpoints shall occur within 10 minutes of opening the operable opening.

Exceptions:

1. Separately zoned areas associated with the preparation of food that contain appliances that contribute to the HVAC loads of a restaurant or similar type of occupancy.
2. Warehouses that utilize overhead doors for the function of the occupancy, where approved by the code official.
3. The first entrance doors where located in the exterior wall and are part of a vestibule system.

C402.5.11.1 Operable controls. Controls shall comply with Section C403.14.

C403.14 Operable opening interlocking controls. The heating and cooling systems shall have controls that will interlock these mechanical systems to the set temperatures of 90°F (32°C) for cooling and 55°F (12.7°C) for heating when the conditions of Section C402.5.11 exist. The controls shall configure to shut off the systems entirely when the outdoor temperatures are below 90°F (32°C) or above 55°F (12.7°C).

Photo courtesy of Cushing Terrell Architects

Operable openings greater than 40 square feet require interlocking.

CHANGE SIGNIFICANCE: A variety of occupancies, such as restaurants, bars and gyms, include large operable windows or roll-up, sliding or folding doors. This design feature is used for both ambience and with the intention of creating open spaces that take advantage of cross ventilation or wind for cooling and ventilation. Without controls, the cooling and heating systems for these spaces are still running, negatively impacting energy efficiency.

This new provision requires openings larger than 40 square feet to be interlocked with the heating and cooling system. The exceptions address very specific situations where this requirement would hinder the function of the space, such as separately zoned areas where food preparation areas contribute to the HVAC loads, warehouse doors and entrance doors that are part of a vestibule system.

The system is time and temperature specific, requiring system adjustments within ten minutes of opening. Required controls raise the cooling setpoint to 90°F (32°C) and lower the heating setpoint to 55°F (12.7°C) whenever the operable opening is open, and shut off the system when outdoor temperatures are between 90°F (32°C) and 55°F (12.7°C).

This new provision is applicable to both IECC compliance paths.

C403

Building Mechanical Systems

Section C403 broadly addresses the efficiency, construction and operation of building mechanical systems. Federal regulation establishes energy efficiency standards for many consumer and commercial appliances under the Energy Policy and Conservation Act (EPCA). States are preempted from creating energy efficiency standards for appliances and equipment that are covered by federal law under the EPCA, 42 U.S.C. 6201, et seq.[2] Thus, the IECC includes both equipment efficiencies set by the U.S. Department of Energy (DOE) under the EPCA and efficiencies for equipment not covered under federal law. Additional efficiencies are derived from insulated and sealed delivery systems and controls.

The 2021 IECC Section C403 is divided into 14 parts:

- Sections C403.1 and C403.2 address general requirements that are applicable to all building cooling, heating and ventilation systems.
- Section C403.3 addresses efficiency requirements for building cooling, heating and ventilation equipment and appliances, turndown requirements for boiler systems and the limitations for hot gas bypass.
- Section C403.4 provides the control requirements for building cooling and heating systems, including hydronic systems controls.
- Section C403.5 addresses both air- and water-side economizer requirements.
- Section C403.6 includes provisions specific to multi-zone systems.
- Section C403.7 applies to building ventilation and exhaust systems.
- Section C403.8 provides efficiency and control requirements for fans.
- Section C403.9 references a standard new to the 2021 IECC for large-diameter ceiling fans.
- Section C403.10 provides requirements for heat rejection equipment.
- Section C403.11 applies to refrigerated warehouse coolers, refrigerated warehouse freezers, walk-in coolers, walk-in freezers and refrigerated display cases.
- Section C403.12 addresses ductwork construction, insulation and testing as well as piping insulation.
- Section C403.13 addresses mechanical systems that provide heat outside the building's thermal envelope.
- Section C403.14 introduces requirements for operable interlocking controls.

Significant changes to the building mechanical systems in the 2021 IECC include:

- Refined HVAC control requirements

---

2 For purposes of this document, EPCA refers the EPCA as amended and codified in 42 U.S.C. §§ 6201-6422. Since its enactment in 1975, EPCA has been amended by the National Energy Conservation Policy Act, Pub. L. No. 95-619, § 422, 92 Stat. 3206 (1978); the National Appliance Energy Conservation Act, Pub. L. No. 100-12, § 7, 101 Stat. 103 (1987); the Energy Policy Act of 1992, Pub. L. 102-486, §§ 121-28, 106 Stat. 2776 (1992); the Energy Policy Act of 2005, Pub. L. 109-58, §§ 135-36, 119 Stat. 594 (2005); Energy Independence and Security Act of 2007, Pub. L. 110-140, §§ 301-325, 121 Stat. 1492 (2007); and the American Energy Manufacturing Technical Corrections Act, Pub. L. 112-210, 126 Stat. 1514 (2012).

Building mechanical systems are addressed in Section C403.

- Added requirements for low-capacity ventilation fans
- Updated fan efficiency requirements

The 2021 IECC also includes two new sections, Section C403.9 Large-diameter ceiling fans, which resulted in subsequent renumbering, and Section C403.14 Operable opening interlocking controls, described in the previous section.

C403.4.2.3

Automatic Start and Stop

CHANGE TYPE: Modification

CHANGE SUMMARY: In addition to automatic start controls, HVAC systems with direct digital controls serving individual zones must have automatic stop controls.

2021 CODE: **C403.4.2.3 Automatic start and stop ~~(Mandatory).~~** Automatic start and stop controls shall be provided for each HVAC system. The automatic start controls shall be configured to automatically adjust the daily start time of the HVAC system in order to bring each space to the desired occupied temperature immediately prior to scheduled occupancy. Automatic stop controls shall be provided for each HVAC system with direct digital control of individual zones. The automatic stop controls shall be configured to reduce the HVAC system's heating temperature setpoint and increase the cooling temperature setpoint by not less than 2°F (−16.6°C) before scheduled unoccupied periods based upon the thermal lag and acceptable drift in space temperature that is within comfort limits.

CHANGE SIGNIFICANCE: The 2018 IECC established requirements for automatic start controls. These controls automatically adjust the start time for warm up or cool down of the zone prior to occupancy so that the zone is at the appropriate temperature when it is scheduled to be occupied. Automatic stop controls, sometimes referred to as an optimum stop, are now required for HVAC systems with direct digital controls serving individual zones. The controls must be capable of decreasing the heating temperature setpoint by 2°F and increasing the cooling setpoint by 2°F for some time period before the scheduled unoccupied period where the temperature can drift within the comfort range using the adjusted setpoints. Typically this results in little heating or cooling for 30 to 60 minutes at the end of the occupied period.

This new provision is applicable to both IECC compliance paths.

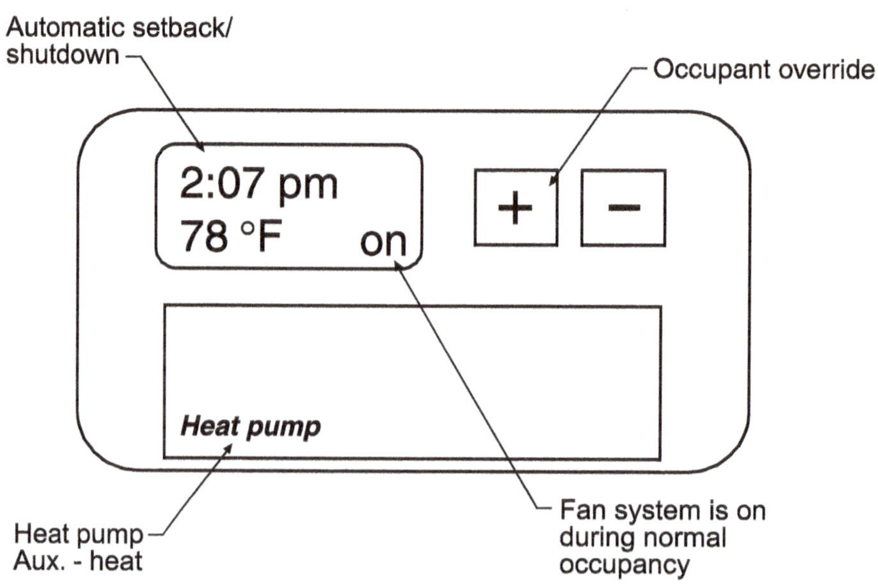

Automatic start and stop are both required.

C403.8.3

Fan Efficiency

CHANGE TYPE: Modification

CHANGE SUMMARY: The fan efficiency metric is updated to the Fan Energy Index, providing consistent fan efficiency requirements across the IECC and ASHRAE 90.1 and resulting in a more effective energy savings metric.

2021 CODE: **C403.8.3 Fan efficiency ~~(Mandatory)~~.** ~~Fans~~ Each fan and fan array shall have a fan ~~efficiency grade~~ energy index ~~(FEG)~~ (FEI) of not less than ~~67~~ 1.00 at the design point of operation, as determined in accordance with AMCA ~~205~~ 208 by an approved independent testing laboratory and labeled by the manufacturer. ~~The total efficiency of the fan~~ Each fan and fan array used for a variable-air-volume system shall have an FEI of not less than 0.95 at the design point of operation, ~~shall be within 15 percentage points of the maximum total efficiency of the fan.~~ as determined in accordance with AMCA 208 by an approved independent testing laboratory and labeled by the manufacturer. The FEI for fan arrays shall be calculated in accordance with AMCA 208 Annex C.

Exceptions: The following fans are not required to have a fan ~~efficiency grade~~: energy index:

1. Fans that are not embedded fans with motor nameplate horsepower of less than 1.0 hp (0.75 kW) or with a nameplate electrical input power of less than 0.89 kW.
2. Embedded fans that have ~~1.Fans of 5 hp (3.7 kW) or less as follows:1.1.Individual fans with~~ a motor nameplate horsepower of 5 hp (3.7 kW) or less, ~~unless Exception 1.2 applies~~. or with a fan system electrical input power of 4.1 kW or less.
3. Multiple fans operated ~~1.2.Multiple fans~~ in series or parallel as the functional equivalent of a single fan that have a combined motor nameplate horsepower of 5 hp (3.7 kW) or less ~~and are operated as the functional equivalent of a single fan~~ or with a fan system electrical input power of 4.1 kW or less.
4. Fans that are part of equipment covered in Section C403.3.2.
5. Fans included in an equipment package certified by an approved agency for air or energy performance.

~~4. Powered wall/roof ventilators.~~

~~5. Fans outside the scope of AMCA 205.~~

6. Ceiling fans, which are defined as nonportable devices suspended from a ceiling or overhead structure for circulating air via the rotation of the blades.
7. Fans used for moving gases at temperatures above 425°F (250°C).
8. Fans used for operation in explosive atmospheres.
9. Reversible fans used for tunnel ventilation.
10. ~~6.~~ Fans that are intended to operate only during emergency conditions.
11. Fans outside the scope of AMCA 208.

CHANGE SIGNIFICANCE: Supported by five new fan-related definitions – embedded fan, fan array, fan nameplate electrical input power, fan system electrical input power and Fan Energy Index (FEI) – this change applies a new fan energy metric, the FEI, to fans and fan arrays. The FEI is discussed in detail on page C-7 of this publication.

Key to the FEI is the reduction in the minimum covered fan size from 5 HP to 1 HP, which results in significant energy savings. Additionally, the FEI removes both the exemption for powered roof ventilators (PRVs) and the "sizing/selection window" of "15 percentage points from peak total efficiency." The inclusion of PRVs results in additional energy savings and the removal of the sizing/selection window makes the provision easier to enforce.

The inclusion of the FEI harmonizes the IECC with ASHRAE 90.1 and anticipated future federal regulations. The FEI was developed in response to the DOE rulemaking for commercial fans and blowers. The rulemaking is currently incomplete; however, updates to ASHRAE 90.1 and the IECC effectively apply the new metric. The FEI has been added to the DOE EnergyPlus software and to the DOE Fan System Assessment Tool.

The FEI does not apply to all fans; smaller fans, those that are part of equipment covered in Section C403.3.2 and other equipment packages, and ceiling fans are exempt, among others.

This new provision is applicable to both IECC compliance paths.

C403.8.5

Low-Capacity Ventilation Fans

CHANGE TYPE: Addition

CHANGE SUMMARY: New requirements for low-capacity ventilation fans apply the efficiencies of ventilation fans typical of residential construction to mid-rise residential occupancies and small commercial buildings.

2021 CODE: **C403.8.5 Low-capacity ventilation fans.** Mechanical ventilation system fans with motors less than 1/12 hp (0.062 kW) in capacity shall meet the efficacy requirements of Table C403.8.5 at one or more rating points.

Exceptions:

1. Where ventilation fans are a component of a listed heating or cooling appliance.
2. Dryer exhaust duct power ventilators, domestic range hoods and domestic range booster fans that operate intermittently.

TABLE C403.8.5 Low-Capacity Ventilation Fan Efficacy[a]

FAN LOCATION	AIRFLOW RATE MINIMUM (CFM)	MINIMUM EFFICACY (CFM/WATT)	AIRFLOW RATE MAXIMUM (CFM)
HRV or ERV	Any	1.2 cfm/watt	Any
In-line fan	Any	3.8 cfm/watt	Any
Bathroom, utility room	10	2.8 cfm/watt	< 90
Bathroom, utility room	90	3.5 cfm/watt	Any

For SI: 1 cfm/ft = 47.82 W.

a. Airflow shall be tested in accordance with HVI 916 and listed. Efficacy shall be listed or shall be derived from listed power and airflow. Fan efficacy for fully ducted HRV, ERV, balanced and in-line fans shall be determined at a static pressure not less than 0.2 inch w.c. Fan efficacy for ducted range hoods, bathroom and utility room fans shall be determined at a static pressure not less than 0.1 inch w.c.

Photo courtesy of Kelly Makela

Low-capacity ventilation fans must meet efficiency requirements.

CHANGE SIGNIFICANCE: Exhaust fan efficacies were introduced into the 2012 IECC for whole-house ventilation in low-rise residential buildings; however, these same requirements were not previously included in the commercial provisions of the IECC despite the application of such fans in some commercial projects. This change applies fan efficiency requirements typical of residential low-rise construction to the same small ventilation fans used in mid-rise residential occupancies and small commercial buildings.

Small ventilation fans are used for point-of-source contaminant exhaust and are frequently incorporated into the ventilation strategy for multiple-family buildings. Such fans are smaller than the threshold for fan size (1/12 hp) that is attached to the other commercial fan requirements.

This change adopts the table approach already utilized for these fans in the residential section of the code and updates the efficiency requirements.

This new provision is applicable to both IECC compliance paths.

C405

Electrical Power and Lighting Systems

Lighting accounts for a significant amount of the energy consumption of a commercial building. Lighting fixtures have seen many advances in technology in recent years and new or improved lighting controls offer opportunities for saving energy without the need to change occupant behavior; together, lighting reductions are automatic and mostly unnoticed. Efficient lighting sources and controls maintain and may even improve interior and exterior illumination. Numerous studies suggest that daylighting improves employee satisfaction with working conditions and reduces absenteeism. Careful attention to lighting design and controls benefits building occupants and building owners.

The provisions in Section C405 apply to lighting systems in new buildings, additions, tenant finishes, alterations to existing lighting systems and any change in occupancy. Controls and minimum lamp efficiencies are required for interior and exterior lighting systems. Exceptions to the provisions include some historic buildings, the replacement of less than 10 percent of the existing interior luminaires and alterations that do not increase the required lighting power.

Significant changes to the electrical power and lighting systems provisions in the 2021 IECC include:

- Expansion and clarification of lighting control applications
- New provisions for plant growth lighting
- New provisions for automatic receptacle control
- New provisions for metering and monitoring

C405.2.1.2

Occupant Sensor Controls in Warehouse Storage Areas

CHANGE TYPE: Clarification

CHANGE SUMMARY: Occupant sensor control functions are updated for warehouse storage areas in Section C405.2.1.2, providing further guidance on how occupant sensors should operate within warehouses.

2021 CODE: **C405.2.1.2 Occupant sensor control function in ~~warehouses~~ <u>warehouse storage areas</u>.** Lighting in warehouse storage areas shall be controlled as follows:

1. Lighting in each aisleway shall be controlled independently of lighting in all other aisleways and open areas.
2. Occupant sensors shall ~~In warehouses, the lighting in aisleways and open areas shall be controlled with occupant sensors that~~ automatically reduce lighting power within each controlled area to an unoccupied setpoint of not more ~~by not less~~ than 50 percent within 20 minutes after all occupants have left the controlled area. ~~when the areas are unoccupied.~~
3. Lights that are not turned off by occupant sensors shall <u>be turned off by time-switch control complying with Section C405.2.2.1</u> ~~control lighting in each aisleway independently and shall not control lighting beyond the aisleway being controlled by the sensor.~~
4. A manual control shall be provided to allow occupants to turn off lights in the space.

CHANGE SIGNIFICANCE: The IECC provisions for occupancy sensor controls continue to evolve, providing further clarification for code users on the application of the technology. Warehouse spaces were added to the occupancy controls list of space types (Section C405.2.1) in the 2015 IECC, yet warehouses were excluded from the specific requirements of the main occupancy sensor control functions. Because of this, and the

Photo courtesy of Kmatta/Moment/Getty Images

Occupant sensor controls are required in warehouse storage areas.

unique control requirements, several considerations for how occupancy sensors should operate in the warehouse environment were overlooked or left out. This created confusion that is resolved with this change.

The revisions do not change the intent of the requirements. Instead, these clarifications reduce inconsistency and eliminate application and enforcement confusion associated with warehouse lighting controls. Similar to the enumerated items put into the 2018 IECC for open office plan spaces, this change enumerates how to apply lighting controls in warehouse storage area spaces as follows:

1. Item one simplifies language on independent control between aisleways and open areas.
2. Item two specifies the time delay or "time out" for occupants leaving the warehouse area and turning off or reducing the lighting is 20 minutes. This aligns with the time delay for all other occupancy sensor spaces (C405.2.1.1) and open-plan office spaces (405.2.1.3), and reduces lighting power.
3. Item three clarifies that the warehouse lighting must automatically turn off according to methods already presented in the code (occupancy sensor or time-switch controls) but leaves the choice of method to the building design professional.
4. Item four clarifies control is available for the end user to turn the lighting off.

This provision is applicable to both IECC compliance paths.

C405.2.8

Parking Garage Lighting Control

CHANGE TYPE: Addition

CHANGE SUMMARY: A new provision for lighting in parking garages recognizes parking garages as a unique space and requires such spaces to have either occupant sensors or time-switch controls.

2021 CODE: **C405.2.8 Parking garage lighting control.** Parking garage lighting shall be controlled by an occupant sensor complying with Section C405.2.1.1 or a time-switch control complying with Section C405.2.2.1. Additional lighting controls shall be provided as follows:

1. Lighting power of each luminaire shall be automatically reduced by not less than 30 percent when there is no activity detected within a lighting zone for 20 minutes. Lighting zones for this requirement shall be not larger than 3,600 square feet (334.5 m^2).

Exception: Lighting zones provided with less than 1.5 footcandles of illumination on the floor at the darkest point with all lights on are not required to have automatic light-reduction controls.

2. Where lighting for eye adaptation is provided at covered vehicle entrances and exits from buildings and parking structures, such lighting shall be separately controlled by a device that automatically reduces lighting power by at least 50 percent from sunset to sunrise.
3. The power to luminaires within 20 feet (6096 mm) of perimeter wall openings shall automatically reduce in response to daylight by at least 50 percent.

Exceptions:

1. Where the opening to-wall-ratio is less than 40 percent as viewed from the interior and encompassing the vertical distance from the driving surface to the lowest structural element.

Photo courtesy of DuKai photographer/Moment/Getty Images

Parking garages must have lighting controls.

2. Where the distance from the opening to any exterior daylight blocking obstruction is less than one-half the height from the bottom of the opening or fenestration to the top of the obstruction.
3. Where openings are obstructed by permanent screens or architectural elements restricting daylight entering the interior space.

CHANGE SIGNIFICANCE: This new control provision clarifies how to apply lighting requirements to parking garages. Previously, code users questioned whether parking garages should be treated as interior spaces, and, if so, how the control requirements applied to an area with different lighting needs than building interior spaces and how the daylight responsive controls of Section C405.2.4 applied.

This new provision recognizes parking garages as a unique space and provides explicit occupant lighting control requirements, daylight responsive control requirements and exceptions for lighting design in these spaces.

This provision is applicable to both IECC compliance paths.

C405.4

Lighting for Plant Growth

CHANGE TYPE: Addition

CHANGE SUMMARY: A provision regulating lighting for plant growth and maintenance has been added to ensure indoor agriculture operations are energy efficient.

2021 CODE: **C405.4 Lighting for plant growth and maintenance.** Not less than 95 percent of the permanently installed luminaires used for plant growth and maintenance shall have a photon efficiency of not less than 1.6 µmol/J as defined in accordance with ANSI/ASABE S640.

CHANGE SIGNIFICANCE: Previous editions of the IECC have not addressed lighting requirements for indoor agriculture plant growth and maintenance, yet indoor growing operations are becoming increasingly common for both traditional farming crops and marijuana. Indoor agriculture energy usage is projected to grow substantially over the next several years due in part to the legalization of medicinal and recreational marijuana growth and sales. The Northwest Power and Conservation Council projects that indoor marijuana growing operations alone will add as much as 300 average megawatts by 2030. This is equivalent to 1.5 percent of total regional electricity demand. Other studies have shown that lighting for indoor agriculture has accounted for 45 percent of the electric load growth in some states.

This new provision and associated standard—ANSI/ASABE S640—require 95 percent of the lighting used for plant growth or maintenance to meet an efficiency metric of 1.6 µmol/J (micromoles per joule). The metric was developed in collaboration with the American Society of Agricultural and Biological Engineers specifically for lighting used for plant growth. It measures the number of photons emitted from the fixture per joule of energy consumed. This minimum photon efficiency is consistent with the Conservation Practice Standard Energy Efficient Lighting System Code 670 issued by the U.S. Department of Agriculture in January 2020.

This provision is applicable to both IECC compliance paths.

Photo courtesy of James O'Neil/The Image Bank/Getty Images

Lighting for plant growth is regulated by the IECC.

C405.11

Automatic Receptacle Control

CHANGE TYPE: Addition

CHANGE SUMMARY: New provisions require at least 50 percent of covered receptacles and 25 percent of branch circuit feeders to be automatically controlled receptacles.

2021 CODE: **C405.11 Automatic receptacle control.** The following shall have automatic receptacle control complying with Section C405.11.1:

1. At least 50 percent of all 125V, 15- and 20-amp receptacles installed in enclosed offices, conference rooms, rooms used primarily for copy or print functions, breakrooms, classrooms and individual workstations, including those installed in modular partitions and module office workstation systems.
2. At least 25 percent of branch circuit feeders installed for modular furniture not shown on the construction documents.

C405.11.1 Automatic receptacle control function. Automatic receptacle controls shall comply with the following:

1. Either split controlled receptacles shall be provided with the top receptacle controlled, or a controlled receptacle shall be located within 12 inches (304.8 mm) of each uncontrolled receptacle.
2. One of the following methods shall be used to provide control:
   2.1. A scheduled basis using a time-of-day operated control device that turns receptacle power off at specific programmed times and can be programmed separately for each day of the week. The control device shall be configured to provide an independent schedule for each portion of the building of not more

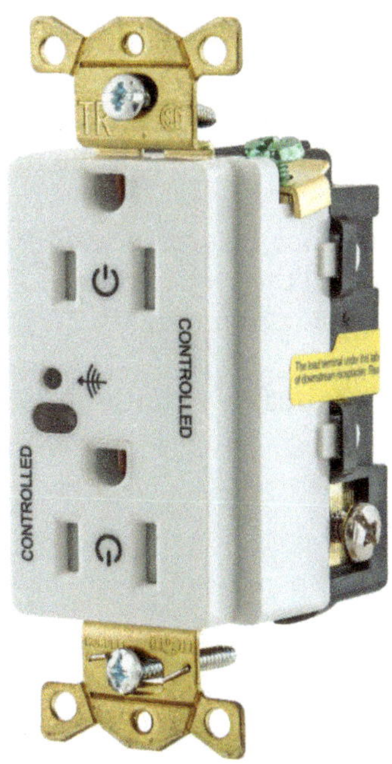

Controlled receptacles are now required.

than 5,000 square feet (464.5 m^2) and not more than one floor. The occupant shall be able to manually override an area for not more than 2 hours. Any individual override switch shall control the receptacles of not more than 5,000 feet (1524 m).

2.2. An occupant sensor control that shall turn off receptacles within 20 minutes of all occupants leaving a space.

2.3. An automated signal from another control or alarm system that shall turn receptacles off within 20 minutes after determining that the area is unoccupied.

3. All controlled receptacles shall be permanently marked in accordance with NFPA 70 and be uniformly distributed throughout the space.

4. Plug-in devices shall not comply.

Exceptions: Automatic receptacles controls are not required for the following:

1. Receptacles specifically designated for equipment requiring continuous operation (24 hours per day, 365 days per year).
2. Spaces where an automatic control would endanger the safety or security of the room or building occupants.
3. Within a single modular office workstation, noncontrolled receptacles are permitted to be located more than 12 inches (304.8 mm), but not more than 72 inches (1828 mm) from the controlled receptacles serving that workstation.

CHANGE SIGNIFICANCE: The inclusion of provisions for controlled receptacles brings practices common outside of North America, and required in California and in ASHRAE 90.1, into the IECC.

At least 50 percent of the covered receptacles and 25 percent of branch circuit feeders must be automatically controlled receptacles. Methods for meeting these requirements include:

1. Using split wired receptacles that provide at least one controlled outlet
2. For any uncontrolled outlets, ensuring that at least one controlled outlet is located within 12 inches of the uncontrolled outlet

To avoid the unintentional loss of power to equipment and appliances, it is critical to note both the requirement – all controlled receptacles be permanently marked in accordance with NFPA 70 – and the application exceptions.

These provisions are applicable to both IECC compliance paths.

C405.12

Energy Monitoring

CHANGE TYPE: Addition

CHANGE SUMMARY: Energy monitoring requirements are added to measure and retain the intended energy performance of a building.

2021 CODE: **C405.12 Energy monitoring.** New buildings with a gross conditioned floor area of 25,000 square feet (2322 m^2) or larger shall be equipped to measure, monitor, record and report energy consumption data in compliance with Sections C405.12.1 through C405.12.5.

> **Exception:** R-2 occupancies and individual tenant spaces are not required to comply with this section provided that the space has its own utility services and meters and has less than 5,000 square feet (464.5 m^2) of conditioned floor area.

C405.12.1 Electrical energy metering. For all electrical energy supplied to the building and its associated site, including but not limited to site lighting, parking, recreational facilities and other areas that serve the building and its occupants, meters or other measurement devices shall be provided to collect energy consumption data for each end-use category required by Section C405.12.2.

C405.12.2 End-use metering categories. Meters or other approved measurement devices shall be provided to collect energy use data for each end-use category indicated in Table 405.12.2. Where multiple meters are used to measure any end-use category, the data acquisition system shall total all of the energy used by that category. Not more than 5 percent of the measured load for each of the end-use categories indicated in Table 405.12.2 shall be permitted to be from a load that is not within that category.

Photo courtesy of Cavan images/Cavan/Getty Images

Energy monitoring supports efficient building operations.

Exceptions:

1. HVAC and water heating equipment serving only an individual dwelling unit shall not require end-use metering.
2. End-use metering shall not be required for fire pumps, stairwell pressurization fans or any system that operates only during testing or emergency.
3. End-use metering shall not be required for an individual tenant space having a floor area not greater than 2,500 square feet (232 m^2) where a dedicated source meter complying with Section C405.12.3 is provided.

TABLE C405.12.2 Energy Use Categories

LOAD CATEGORY	DESCRIPTION OF ENERGY USE
Total HVAC system	Heating, cooling and ventilation, including but not limited to fans, pumps, boilers, chillers and water heating. Energy used by 120-volt equipment, or by 208/120-volt equipment that is located in a building where the main service is 480/277-volt power, is permitted to be excluded from total HVAC system energy use.
Interior lighting	Lighting systems located within the building.
Exterior lighting	Lighting systems located on the building site but not within the building.
Plug loads	Devices, appliances and equipment connected to convenience receptacle outlets.
Process loads	Any single load that is not included in a HVAC, lighting or plug load category and that exceeds 5 percent of the peak connected load of the whole building, including but not limited to data centers, manufacturing equipment and commercial kitchens.
Building operations and other miscellaneous loads	The remaining loads not included elsewhere in this table, including but not limited to vertical transportation systems, automatic doors, motorized shading systems, ornamental fountains, ornamental fireplaces, swimming pools, in-ground spas and snow-melt systems.

C405.12.3 Meters. Meters or other measurement devices required by this section shall be configured to automatically communicate energy consumption data to the data acquisition system required by Section C405.12.4. Source meters shall be allowed to be any digital-type meter. Lighting, HVAC or other building systems that can monitor their energy consumption shall be permitted instead of meters. Current sensors shall be permitted, provided that they have a tested accuracy of ± 2 percent. Required metering systems and equipment shall have the capability to provide at least hourly data that is fully integrated into the data acquisition system and graphical energy report in accordance with Sections C405.12.4 and C405.12.5.

C405.12.4 Data acquisition system. A data acquisition system shall have the capability to store the data from the required meters and other sensing devices for a minimum of 36 months. The data acquisition system shall have the capability to store real-time energy consumption data and provide hourly, daily, monthly and yearly logged data for each end-use category required by Section C405.12.2.

C405.12.5 Graphical energy report. A permanent and readily accessible reporting mechanism shall be provided in the building that is accessible by building operation and management personnel. The reporting mechanism shall have the capability to graphically provide the energy consumption for each end-use category required by Section C405.12.2 at least every hour, day, month and year for the previous 36 months.

CHANGE SIGNIFICANCE: The 2021 IECC includes new requirements for energy monitoring. Energy monitoring provides a means to measure and retain the energy performance of a building. Although the IECC requires mechanical system and service water system commissioning, without a monitoring system in place, building owners and operators often do not know if energy efficiency measures have degraded, gone out of calibration, or are experiencing some other issue that is reducing the return on investment and increasing the payback period.

Electricity metering and monitoring is a management tool that uses a combination of hardware and software to gather data and turn them into useful and actionable information about a building, system or specific equipment. The requirements in this new section are intended to save energy by continually monitoring and reporting actionable energy consumption data to building owners and operators.

Monitoring subsystems provides key indications when changes have been made or systems are not operating to specification, which increases energy consumption. Examples of issues identified through monitoring include these:

1. Increased energy consumption in HVAC system loads may indicate failures in motors, drive systems, bearings or other system components.
2. Increased energy consumption from lighting loads may indicate there has been a failure of occupant sensors, lighting schedules are not working appropriately, lamps need to be replaced or cleaned or building occupants have installed additional lighting.
3. Increased energy consumption of the entire building may indicate the building envelope is degrading or configuration changes have been made that impact total building energy use.
4. Increased energy consumption of plug loads may indicate when computer equipment is left on during nonworking hours or when space heaters are used that compromise the efficiency of the facility due to setpoints on the HVAC system.

These new provisions are applicable to both IECC compliance paths.

C406

Additional Efficiency Requirements

First introduced in the 2012 IECC, the additional efficiency provisions of Section C406 are required for commercial buildings using the Prescriptive Compliance option. The 2021 IECC includes a substantial revision to this section, increasing the number of options, introducing a points-based system and providing a climate zone-specific point value for each option.

The additional efficiency options do not take the building above the minimum requirements of the energy code, but instead ensure buildings complying prescriptively achieve the same level of efficiency as the Total Building Performance option. This provides flexibility for the building owner and designer to decide how best to make their building or system more efficient, keeping with the IECC's original intent of providing flexibility in order to conserve energy in buildings.

C406

Additional Efficiency Requirements

CHANGE TYPE: Modification

CHANGE SUMMARY: When complying with the Prescriptive Compliance option, designers may select from 11 points-based additional energy efficiency options that now consider climate zone and occupancy type.

2021 CODE: **C406.1 Additional energy efficiency credit requirements.** ~~Buildings shall comply~~ New buildings shall achieve a total of 10 credits from Tables C406.1(1) through C406.1(5) where the table is selected based on the use group of the building and from credit calculations as specified in relevant subsections of Section C406. Where a building contains multiple-use groups, credits from each use group shall be weighted by floor area of each group to determine the weighted average building credit. Credits from the tables or calculation shall be achieved where a building complies with one or more of the following:

1. More efficient HVAC performance in accordance with Section C406.2.
2. Reduced lighting power in accordance with Section C406.3.
3. Enhanced lighting controls in accordance with Section C406.4.
4. On-site supply of renewable energy in accordance with Section C406.5.
5. Provision of a dedicated outdoor air system for certain HVAC equipment in accordance with Section C406.6.
6. High-efficiency service water heating in accordance with Section C406.7.
7. Enhanced envelope performance in accordance with Section C406.8.
8. Reduced air infiltration in accordance with Section C406.9
9. Where not required by Section C405.12, include an energy monitoring system in accordance with C406.10.

Photo courtesy of roman023/iStock/Getty Images Plus

Additional energy efficiency requirements now consider climate and occupancy type.

10. Where not required by Section C403.2.3, include a fault detection and diagnostics (FDD) system in accordance with Section C406.11.
11. Efficient kitchen equipment in accordance with Section C406.12.

C406.1.1 Tenant spaces. Tenant spaces shall comply with sufficient options from Tables C406.1(1) through C406.1(5) to achieve a minimum number of 5 credits, where credits are selected from Section C406.2, C406.3, C406.4, C406.6, C406.7 or C406.10. ~~Alternatively, tenant spaces shall comply with Section C406.5 where the entire building is in compliance~~. Where the entire building complies using credits from Section C406.5, C406.8 or C406.9, tenant spaces shall be deemed to comply with this section.

Exception: Previously occupied tenant spaces that comply with this code in accordance with Section C501.

TABLE C406.1(1) Additional Energy Efficiency Credits for Group B Occupancies

	Climate Zone																
Section	0A & 1A	0B & 1B	2A	2B	3A	3B	3C	4A	4B	4C	5A	5B	5C	6 A	6 B	7	8
C406.2.1: 5% heating efficiency improvement	NA	NA	NA	NA	NA	NA	NA	NA	NA	NA	1	NA	NA	1	1	NA	1
C406.2.2: 5% cooling efficiency improvement	6	6	5	5	4	4	3	3	3	2	2	2	1	2	2	2	1
C406.2.3: 10% heating efficiency improvement	NA	NA	NA	NA	NA	NA	NA	NA	NA	NA	2	1	1	2	2	NA	1
C406.2.4: 10% cooling efficiency improvement	11	12	10	9	7	7	6	5	6	4	4	5	3	4	3	3	3
C406.3: Reduced lighting power	9	8	9	9	9	9	10	8	9	9	7	8	8	6	7	7	6
C406.4: Enhanced digital lighting controls	2	2	2	2	2	2	2	2	2	2	2	2	2	1	2	1	1
C406.5: On-site renewable energy	9	9	9	9	9	9	9	9	9	9	9	9	9	9	9	9	9
C406.6: Dedicated outdoor air	4	4	4	4	4	3	2	5	3	2	5	3	2	7	4	5	3
C406.7.2: Recovered or renewable water heating	NA	NA	NA	NA	NA	NA	NA	NA	NA	NA	NA	NA	NA	NA	NA	NA	NA
C406.7.3: Efficient fossil fuel water heater	NA	NA	NA	NA	NA	NA	NA	NA	NA	NA	NA	NA	NA	NA	NA	NA	NA
C406.7.4: Heat pump water heater	NA	NA	NA	NA	NA	NA	NA	NA	NA	NA	NA	NA	NA	NA	NA	NA	NA
C406.8: Enhanced envelope performance	1	4	2	4	4	3	NA	7	4	5	10	7	6	11	10	14	16
C406.9: Reduced air infiltration	2	1	1	2	4	1	NA	8	2	3	11	4	1	15	8	11	6
C406.10: Energy monitoring	4	4	4	4	3	3	3	3	3	3	2	3	2	2	2	2	2
C406.11: Fault detection and diagnostics system	2	2	2	2	1	1	1	1	1	1	1	1	1	1	1	1	1

NA=Not Applicable.

TABLE C406.1(2) Additional Energy Efficiency Credits for Group R and I Occupancies

Section	Climate Zone																
	0A & 1A	0B & 1B	2A	2B	3A	3B	3C	4A	4B	4C	5A	5B	5C	6 A	6 B	7	8
C406.2.1: 5% heating efficiency improvement	NA	NA	NA	NA	1	NA	NA	1	NA	1	1	1	1	2	1	2	2
C406.2.2: 5% cooling efficiency improvement	3	3	2	2	1	1	1	1	1	NA	1	1	NA	1	1	1	NA
C406.2.3: 10% heating efficiency improvement	NA	NA	NA	NA	1	NA	NA	1	1	1	2	2	1	3	2	3	4
C406.2.4: 10% cooling efficiency improvement	5	5	4	3	2	3	1	2	2	1	1	1	1	1	1	1	1
C406.3: Reduced lighting power	2	2	2	2	2	2	3	2	2	2	2	2	2	2	2	2	2
C406.4: Enhanced digital lighting controls	NA	NA	NA	NA	NA	NA	NA	NA	NA	NA	NA	NA	NA	NA	NA	NA	NA
C406.5: On-site renewable energy	8	8	8	8	7	8	8	7	7	7	7	7	7	7	7	7	7
C406.6: Dedicated outdoor air	3	4	3	3	4	2	NA	6	3	4	8	5	5	10	7	11	12
C406.7.2: Recovered or renewable water heating	10	9	11	10	13	12	15	14	14	15	14	14	16	14	15	15	15
C406.7.3: Efficient fossil fuel water heater	5	5	6	6	8	7	8	8	8	9	9	9	10	10	9	10	11
C406.7.4: Heat pump water heater	6	5	5	5	5	5	5	5	5	5	5	5	5	5	5	5	5
C406.8: Enhanced envelope performance	3	6	3	5	4	4	1	4	3	3	4	5	3	5	4	6	6
C406.9: Reduced air infiltration	6	5	3	11	6	4	NA	7	3	3	9	5	1	13	6	8	3
C406.10: Energy monitoring	1	1	1	1	1	1	1	1	1	1	1	1	1	1	1	1	1
C406.11: Fault detection and diagnostics system	1	1	1	1	1	1	NA	1	1	NA	1	1	NA	1	1	1	1

NA=Not Applicable.

TABLE C406.1(3) Additional Energy Efficiency Credits for Group E Occupancies

Section	Climate Zone																
	0A & 1A	0B & 1B	2A	2B	3A	3B	3C	4A	4B	4C	5A	5B	5C	6 A	6 B	7	8
C406.2.1: 5% heating efficiency improvement	NA	NA	NA	NA	1	1	1	1	1	2	1	2	1	2	2	3	4
C406.2.2: 5% cooling efficiency improvement	4	4	3	3	2	2	2	2	1	1	1	1	NA	1	1	1	NA
C406.2.3: 10% heating efficiency improvement	NA	NA	NA	1	1	1	1	2	3	4	3	4	3	4	3	5	7
C406.2.4: 10% cooling efficiency improvement	7	8	7	6	5	4	3	4	3	1	2	2	1	2	2	2	1
C406.3: Reduced lighting power	8	8	8	9	8	9	9	8	9	9	8	9	8	7	8	7	7
C406.4: Enhanced digital lighting controls	2	2	2	2	2	2	2	2	2	2	2	2	2	2	2	2	1
C406.5: On-site renewable energy	6	6	6	6	6	6	6	6	6	6	6	6	6	6	6	5	5

(Continued)

TABLE C406.1(3) (*Continued*)

	Climate Zone																
Section	0A & 1A	0B & 1B	2A	2B	3A	3B	3C	4A	4B	4C	5A	5B	5C	6 A	6 B	7	8
C406.6: Dedicated outdoor air	NA	NA	NA	NA	NA	NA	NA	NA	NA	NA	NA	NA	NA	NA	NA	NA	NA
C406.7.2: Recovered or renewable water heating[a]	1	1	1	1	1	1	1	1	1	1	1	1	1	1	1	1	1
C406.7.3: Efficient fossil fuel water heater[a]	NA	1	1	1	1	1	1	2	2	3	2	3	2	3	3	3	5
C406.7.4: Heat pump water heater[a]	NA	NA	NA	NA	NA	NA	NA	1	NA	NA	1	1	NA	1	1	1	1
C406.8: Enhanced envelope performance	3	7	3	4	2	4	1	1	3	1	2	3	NA	4	3	6	9
C406.9: Reduced air infiltration	1	1	1	2	NA	NA	NA	NA	NA	NA	1	NA	NA	4	1	4	3
C406.10: Energy monitoring	3	3	3	3	3	3	3	3	3	2	2	3	2	2	2	2	2
C406.11: Fault detection and diagnostics system	1	2	1	1	1	1	1	1	1	1	1	1	1	1	1	1	2

NA=Not Applicable.
a. For schools with showers or full-service kitchens.

TABLE C406.1(4) Additional Energy Efficiency Credits for Group M Occupancies

	Climate Zone																
Section	0A & 1A	0B & 1B	2A	2B	3A	3B	3C	4A	4B	4C	5A	5B	5C	6 A	6 B	7	8
C406.2.1: 5% heating efficiency improvement	NA	NA	NA	NA	1	1	NA	1	1	2	2	2	2	3	2	3	4
C406.2.2: 5% cooling efficiency improvement	5	6	4	4	3	3	1	2	2	1	1	2	NA	1	1	1	NA
C406.2.3: 10% heating efficiency improvement	NA	NA	NA	1	1	1	1	2	2	4	3	4	5	5	3	6	8
C406.2.4: 10% cooling efficiency improvement	9	12	9	8	6	6	3	4	4	1	2	3	NA	2	2	2	1
C406.3: Reduced lighting power	13	13	15	14	16	14	17	15	15	14	12	14	14	16	16	14	12
C406.4: Enhanced digital lighting controls	3	3	4	3	4	3	4	4	4	3	3	3	3	4	4	3	3
C406.5: On-site renewable energy	8	8	8	8	8	8	8	8	8	7	7	7	7	7	7	7	6
C406.6 : Dedicated outdoor air	3	4	3	3	3	3	1	3	2	2	2	3	2	4	3	4	4
C406.7.2: Recovered or renewable water heating	NA	NA	NA	NA	NA	NA	NA	NA	NA	NA	NA	NA	NA	NA	NA	NA	NA
C406.7.3: Efficient fossil fuel water heater	NA	NA	NA	NA	NA	NA	NA	NA	NA	NA	NA	NA	NA	NA	NA	NA	NA
C406.7.4: Heat pump water heater	NA	NA	NA	NA	NA	NA	NA	NA	NA	NA	NA	NA	NA	NA	NA	NA	NA
C406.8: Enhanced envelope performance	4	6	3	4	3	3	1	6	4	4	4	5	4	6	5	8	9
C406.9: Reduced air infiltration	1	1	1	2	1	1	NA	3	1	1	3	2	1	7	3	6	3
C406.10: Energy monitoring	4	5	5	5	5	4	4	4	4	3	3	4	3	4	4	4	3
C406.11: Fault detection and diagnostics system	2	2	2	2	1	1	1	1	1	1	1	1	1	1	1	2	2

TABLE C406.1(5) Additional Energy Efficiency Credits for Other[a] Occupancies

Section	Climate Zone																
	0A & 1A	0B & 1B	2A	2B	3A	3B	3C	4A	4B	4C	5A	5B	5C	6 A	6 B	7	8
C406.2.1: 5% heating efficiency improvement	NA	NA	NA	NA	1	1	1	1	1	2	1	2	1	2	2	3	3
C406.2.2: 5% cooling efficiency improvement	5	5	4	4	3	3	2	2	2	1	1	2	1	1	1	1	1
C406.2.3: 10% heating efficiency improvement	NA	NA	NA	1	1	1	1	2	2	3	3	3	3	4	3	5	5
C406.2.4: 10% cooling efficiency improvement	8	9	8	7	5	5	3	4	4	2	2	3	2	2	2	2	2
C406.3: Reduced lighting power	8	8	9	9	9	9	10	8	9	9	7	8	8	8	8	8	7
C406.4: Enhanced digital lighting controls	2	2	2	2	2	2	2	2	2	2	2	3	2	2	2	2	1
C406.5: On-site renewable energy	8	8	8	8	8	8	8	8	8	7	7	7	7	7	7	7	7
C406.6: Dedicated outdoor air	3	4	3	3	4	3	2	5	3	3	5	4	3	7	5	7	6
C406.7.2: Recovered or renewable water heating[b]	10	9	11	10	13	12	15	14	14	15	14	14	16	14	15	15	15
C406.7.3: Efficient fossil fuel water heater[b]	5	5	6	6	8	7	8	8	8	9	9	9	10	10	9	10	11
C406.7.4: Heat pump water heater[b]	6	5	5	5	5	5	5	5	5	5	5	5	5	5	5	5	5
C406.8: Enhanced envelope performance	3	6	3	4	3	4	1	5	4	3	5	5	4	7	6	9	10
C406.9: Reduced air infiltration	3	2	2	4	4	2	NA	6	2	2	6	4	1	10	5	7	4
C406.10: Energy monitoring	3	3	3	3	3	3	3	3	3	3	2	3	2	2	2	3	2
C406.11: Fault detection and diagnostics system	2	2	2	2	1	1	1	1	1	1	1	1	1	1	1	1	1

NA=Not Applicable.
a. Other occupancy groups include all groups except for Groups B, E, I, M and R.
b. For occupancy groups listed in C406.7.1.

C406.2 More efficient HVAC equipment performance. Equipment shall exceed the minimum efficiency requirements listed in ~~Tables C403.3.2(1) through C403.3.2(7) by~~ the tables in Section C403.3.2. ~~10 percent, in addition to the requirements of Section C403. Where multiple performance requirements are provided, the equipment shall exceed all requirements by 10 percent.~~ Variable refrigerant flow systems ~~shall exceed~~ listed in the energy efficiency provisions of ANSI/ASHRAE/IES 90.1 ~~by 10 percent.~~ in accordance with Sections C406.2.1, C406.2.2, C406.2.3 or C406.2.4 shall also meet applicable requirements of Section C403. Energy efficiency credits for heating shall be selected from C406.2.1 or C406.2.3 and energy efficiency credits for cooling shall be selected from C406.2.2, C406.2.4 or C406.2.5. Selected credits shall include a heating or cooling energy efficiency credit or both. Equipment not listed in Tables C403.3.2(1) through C403.3.2~~(7)~~ (9) and variable refrigerant flow systems not listed in the energy efficiency provisions of ASHRAE/IES 90.1 shall be limited to 10 percent of the total building system capacity for heating equipment where selecting Section C406.2.1 or C406.2.3 and cooling equipment where selecting Section C406.2.2, C406.2.4 or C406.2.5.

C406.2.1 Five-percent heating efficiency improvement. Equipment shall exceed the minimum heating efficiency requirements by 5 percent.

C406.2.2 Five-percent cooling efficiency improvement. Equipment shall exceed the minimum cooling and heat rejection efficiency requirements by 5 percent. Where multiple cooling performance requirements are provided, the equipment shall exceed the annual energy requirement, including IEER, SEER and IPLV.

C406.2.3 Ten-percent heating efficiency improvement. Equipment shall exceed the minimum heating efficiency requirements by 10 percent.

C406.2.4 Ten-percent cooling efficiency improvement. Equipment shall exceed the minimum cooling and heat rejection efficiency requirements by 10 percent. Where multiple cooling performance requirements are provided, the equipment shall exceed the annual energy requirement, including IEER, SEER and IPLV.

C406.2.5 More than ten percent cooling efficiency improvement. Where equipment exceeds the minimum annual cooling and heat rejection efficiency requirements by more than 10 percent, energy efficiency credits for cooling may be determined using Equation 4-12, rounded to the nearest whole number. Where multiple cooling performance requirements are provided, the equipment shall exceed the annual energy requirement, including IEER, SEER, and IPLV.

$$EEC_{HEC} = EEC_{10} \times [1 + ((CEI - 10\text{ percent}) \div 10\text{ percent})]$$

(Equation 4-12)

Where:

EEC_{HEC} = Energy efficiency credits for cooling efficiency improvement.

EEC_{10} = Section C406.2.4 credits from Tables C406.1(1) through C406.1(5).

CEI = The lesser of: the improvement above minimum cooling and heat rejection efficiency requirements or 15 percent.

C406.3 Reduced lighting power. Buildings shall comply with Section C406.3.1 or C406.3.2, and dwelling units and sleeping units within the building shall comply with C406.3.3.

C406.3.1 Reduced lighting power by more than ten percent. The total connected interior lighting power calculated in accordance with Section C405.3.1 shall be less than 90 percent of the total lighting power allowance calculated in accordance with Section C405.3.2.

C406.3.2 Reduced lighting power by more than 15 percent. Where the total connected interior lighting power calculated in accordance with Section C405.3.1 is less than 85 percent of the total lighting power allowance calculated in accordance with Section C405.3.2, additional energy efficiency credits shall be determined based on Equation 4-13, rounded to the nearest whole number.

$AEEC_{LPA} = AEEC_{10} \times 10 \times (LPA - LPD) / LPA$ **(Equation 4-13)**

Where:

$AEEC_{LPA}$ = Section C406.3.2 additional energy efficiency credits.
$AEEC_{10}$ = Section C406.3.1 credits from Tables C406.1(1) through C406.1(5).
LPA = Total lighting power allowance calculated in accordance with Section C405.3.2.
LPD = Total connected interior lighting power calculated in accordance with Section C405.3.1.

C406.3.3 Lamp efficacy. Not less than 95 percent of the permanently installed lighting, excluding kitchen appliance light fixtures, serving dwelling units and sleeping units shall be provided by lamps with an efficacy of not less than 65 lumens per watt or luminaires with an efficacy of not less than 45 lumens per watt.

C406.5 On-site renewable energy. Buildings shall comply with Section C406.5.1 or C406.5.2.~~The total minimum ratings of on-site renewable energy systems shall be one of the following:~~

1. ~~Not less than .86 Btu/h per square foot (5.4 W/m^2) or .25 watts per square foot (2.7 W/m^2) of conditioned floor area.~~
2. ~~Not less than 2 percent of the energy used within the building for building mechanical and service water heating equipment and lighting regulated in Chapter 4.~~

C406.5.1 Basic renewable credit. The total minimum ratings of on-site renewable energy systems, not including systems used for credits under Section C406.7.2, shall be one of the following:

1. Not less than 0.86 Btu/h per square foot (2.7 W/m^2) or 0.25 watts per square foot (2.7 W/m^2) of conditioned floor area.
2. Not less than 2 percent of the annual energy used within the building for building mechanical and service water-heating equipment and lighting regulated in Section C405.

C406.5.2 Enhanced renewable credit. Where the total minimum ratings of on-site renewable energy systems exceeds the rating in Section C406.5.1, additional energy efficiency credits shall be determined based on Equation 4-14, rounded to the nearest whole number.

$AEEC_{RRa} = AEEC_{2.5} \times RRa / RR_1$ **(Equation 4-14)**

Where:

$AEEC_{RRa}$ = Section C406.5.2 additional energy efficiency credits.
$AEEC_{2.5}$ = Section C406.5 credits from Tables C406.1(1) through C406.1(5).
RRa = Actual total minimum ratings of on-site renewable energy systems (in Btu/h, watts per square foot or W/m^2).
RR1 = Minimum ratings of on-site renewable energy systems required by Section C406.5.1 (in Btu/h, watts per square foot or W/m^2).

C406.7.~~1~~2 ~~Load fraction.~~ Recovered or renewable water heating. The building service water-heating system shall have one or more of the following that are sized to provide not less than 30 ~~60~~ percent of the building's annual hot water requirements, or sized to provide ~~100~~ 70 percent of the building's annual hot water requirements if the building ~~shall otherwise~~ is required to comply with Section C403. 10~~9~~.5:

1. Waste heat recovery from service hot water, heat-recovery chillers, building equipment or process equipment.
2. On-site renewable energy water-heating systems.

C406.7.3 Efficient fossil fuel water heater. The combined input-capacity weighted-average equipment rating of all fossil fuel water-heating equipment in the building shall be not less than 95 percent Et or 0.95 EF. This option shall receive only half the listed credits for buildings required to comply with C404.2.1.

C406.7.4 Heat pump water heater. Where electric resistance water heaters are allowed, all service hot water system heating requirements shall be met using heat pump technology with a combined input-capacity weighted-average EF of 3.0. Air-source heat pump water heaters shall not draw conditioned air from within the building, except exhaust air that would otherwise be exhausted to the exterior.

C406.10 Energy Monitoring. Buildings shall be equipped to measure, monitor, record and report energy consumption data in compliance with Sections C406.10.1 through C406.10.5.

C406.10.1 Electrical energy metering. For all electrical energy supplied to the building and its associated site, including but not limited to site lighting, parking, recreational facilities, and other areas that serve the building and its occupants, meters or other measurement devices shall be provided to collect energy consumption data for each end-use category required by Section C406.10.2.

C406.10.2 End-use metering categories. Meters or other approved measurement devices shall be provided to collect energy use data for each end-use category listed in Table 406.10.2. These meters shall have the capability to collect energy consumption data for the whole building or for each separately metered portion of the building. Where multiple meters are used to measure any end-use category, the data acquisition system shall total all of the energy used by that category. Not more than 5 percent of the measured load for each of the end-use categories listed in Table 406.10.2 is permitted to be from a load not within the category.

Exceptions:

1. HVAC and water heating equipment serving only an individual dwelling unit does not require end-use metering.
2. End-use metering is not required for fire pumps, stairwell pressurization fans or any system that operates only during testing or emergency.

TABLE 406.10.2 Energy Use Categories

LOAD CATEGORY	DESCRIPTION OF ENERGY USE
Total HVAC system	Heating, cooling and ventilation, including but not limited to fans, pumps, boilers, chillers and water heating. Energy used by 120-volt equipment, or by 208/120-volt equipment that is located in a building where the main service is 480/277-volt power, is permitted to be excluded from total HVAC system energy use.
Interior lighting	Lighting systems located within the building.
Exterior lighting	Lighting systems located on the building site but not within the building.
Plug loads	Devices, appliances and equipment connected to convenience receptacle outlets.
Process loads	Any single load that is not included in an HVAC, lighting or plug load category and that exceeds 5 percent of the peak connected load of the whole building, including but not limited to data centers, manufacturing equipment and commercial kitchens.
Building operations and other miscellaneous loads	The remaining loads not included elsewhere in this table, including but not limited to vertical transportation systems and automatic doors.

C406.10.3 Meters. Meters or other measurement devices required by this section shall be configured to automatically communicate energy consumption data to the data acquisition system required by Section C406.10.4. Source meters shall be allowed to be any digital-type meter. Lighting, HVAC or other building systems that can monitor their energy consumption shall be permitted instead of meters. Current sensors shall be permitted, provided that they have a tested accuracy of ±2 percent. Required metering systems and equipment shall have the capability to provide at least hourly data that is fully integrated into the data acquisition system and graphical energy report in accordance with Sections 406.10.4 and C406.10.5.

C406.10.4 Data acquisition system. A data acquisition system shall have the capability to store the data from the required meters and other sensing devices for a minimum of 36 months. The data acquisition system shall have the capability to store real-time energy consumption data and provide hourly, daily, monthly and yearly logged data for each end-use category required by Section C406.10.2.

C406.10.5 Graphical energy report. A permanent and readily accessible reporting mechanism shall be provided in the building that is accessible by building operation and management personnel. The reporting mechanism shall have the capability to graphically provide the energy consumption for each end-use category required by Section C406.10.2 at least every hour, day, month and year for the previous 36 months.

C406.11 Fault detection and diagnostics system. A fault detection and diagnostics system shall be installed to monitor the HVAC system's performance and automatically identify faults. The system shall do all of the following:

1. Include permanently installed sensors and devices to monitor the HVAC system's performance.

2. Sample the HVAC system performance at least once every 15 minutes.
3. Automatically identify and report HVAC system faults.
4. Automatically notify authorized personnel of identified HVAC system faults.
5. Automatically provide prioritized recommendations for repair of identified faults based on analysis of data collected from the sampling of the HVAC system performance.
6. Be capable of transmitting the prioritized fault repair recommendations to remotely located authorized personnel.

C406.12 Efficient kitchen equipment. For buildings and spaces designated as Group A-2 or facilities that include a commercial kitchen with at least one gas or electric fryer, all fryers, dishwashers, steam cookers and ovens shall comply with all of the following:

1. Achieve performance levels in accordance with the equipment specifications listed in Tables C406.12(1) through C406.12(4) when rated in accordance with the applicable test procedure.
2. Be installed prior to the issuance of the Certificate of Occupancy.
3. Have associated performance levels listed on the construction documents submitted for permitting.

Energy efficiency credits for efficient kitchen equipment shall be independent of climate zone and determined based on Equation 4-15, rounded to the nearest whole number.

$$AEEC_K = 20 \times Area_K / Area_B \qquad \textbf{(Equation 4-15)}$$

Where:

$AEEC_K$ = Section C406.10 additional energy efficiency credits.
$Area_K$ = Floor area of full-service kitchen (ft^2 or m^2).
$Area_B$ = Gross floor area of building (ft^2 or m^2).

TABLE C406.12(1) Minimum Efficiency Requirements: Commercial Fryers

Fryer Type	Heavy-Load Cooking Energy Efficiency	Idle Energy Rate	Test Procedure
Standard open deep-fat gas fryers	≥ 50%	≤ 9,000 Btu/h	ASTM F1361
Standard open deep-fat electric fryers	≥ 83%	≤ 800 watts	
Large-vat open deep-fat gas fryers	≥ 50%	≤ 12,000 Btu/h	ASTM F2144
Large-vat open deep-fat electric fryers	≥ 80%	≤ 1,100 watts	

For SI: 1 Btu/h = 0.293/W.

TABLE C406.12(2) Minimum Efficiency Requirements: Commercial Steam Cookers

Fuel Type	Pan Capacity	Cooking Energy Efficiency[a]	Idle Energy Rate	Test Procedure
Electric steam	3-pan	50%	400 watts	ASTM F1484
	4-pan	50%	530 watts	
	5-pan	50%	670 watts	
	6-pan and larger	50%	800 watts	
Gas steam	3-pan	38%	6,250 Btu/h	
	4-pan	38%	8,350 Btu/h	
	5-pan	38%	10,400 Btu/h	
	6-pan and larger	38%	12,500 Btu/h	

a. Cooking energy efficiency is based on heavy load (potato) cooking capacity.

TABLE C406.12(3) Minimum Efficiency Requirements: Commercial Dishwashers

Machine Type	High-Temperature Efficiency Requirements		Low-Temperature Efficiency Requirements		Test Procedure
	Idle Energy Rate[a]	Water Consumption[b]	Idle Energy Rate[a]	Water Consumption[b]	–
Under counter	≤ 0.50 kW	≤ 0.86 GPR	≤ 0.50 kW	≤ 1.19 GPR	ASTM F1696 ASTM F1920
Stationary single-tank door	≤ 0.70 kW	≤ 0.89 GPR	≤ 0.60 kW	≤ 1.18 GPR	
Pot, pan and utensil	≤ 1.20 kW	≤ 0.58 GPR	≤ 1.00 kW	≤ 0.58 GPSF	
Single-tank conveyor	≤ 1.50 kW	≤ 0.70 GPR	≤ 1.50 kW	≤ 0.79 GPR	
Multiple-tank conveyor	≤ 2.25 kW	≤ 0.54 GPR	≤ 2.00 kW	≤ 0.54 GPR	
Single-tank flight type	Reported	GPH ≤ 2.975x + 55.00	Reported	GPH ≤ 2.975x + 55.00	
Multiple-tank flight type	Reported	GPH ≤ 4.96x + 17.00	Reported	GPH ≤ 4.96x + 17.00	

a. Idle results shall be measured with the door closed and represent the total idle energy consumed by the machine, including all tank heaters and controls. Booster heater (internal or external) energy consumption shall not be part of this measurement unless it cannot be separately monitored.

b. GPR = gallons per rack; GPSF = gallons per square foot of rack; GPH = gallons per hour; x = maximum conveyor belt speed (feet/minute) × conveyor belt width (feet).

TABLE C406.12(4) Minimum Efficiency Requirements: Commercial Ovens

Fuel Type	Classification	Idle Rate	Cooking-Energy Efficiency, %	Test Procedure
		Convection Ovens		
Gas	Full-Size	≤ 12,000 Btu/h	≥ 46	ASTM F1496
Electric	Half-Size	≤ 1.0 Btu/h	≥ 71	
	Full-Size	≤ 1.60 Btu/h		
		Combination Ovens		
Gas	Steam Mode	≤ 200P[a]+6,511 Btu/h	≥ 41	ASTM F2861
	Convection Mode	≤ 150P[a]+5,425 Btu/h	≥ 56	
Electric	Steam Mode	≤ 0.133P[a]+0.6400 kW	≥ 55	
	Convection Mode	≤ 0.080P[a]+0.4989 kW	≥ 76	
		Rack Ovens		
Gas	Single	≤ 25,000 Btu/h	≥ 48	ASTM F2093
	Double	≤ 30,000 Btu/h	≥ 52	

For SI: 1 Btu/h = 0.293/W.
a. P = Pan Capacity: the number of steam table pans the combination oven is able to accommodate in accordance with ASTM F1495.

CHANGE SIGNIFICANCE: The additional efficiency options were introduced into the 2012 IECC as part of the Prescriptive Compliance path to achieve an additional 4 percent energy savings over the base prescriptive requirements of the code. The original language included three additional efficiency options: reduced lighting power density, increased HVAC efficiency and renewable energy. The options were expanded over the next iterations of the code, with eight equal-weighted options to select from included in the 2018 IECC. PNNL performed an analysis to determine the energy savings potential for each of the eight options by climate zone and found significant savings differences. To address this issue, PNNL developed a points-based option that ensures equity across the efficiency measure options, providing the basis for revisions to Section C406.

A total of ten credits (also referred to as points) are now required from Tables C406.1(1) through C406.1(5) and/or efficient kitchen equipment points calculated in accordance with Section C406.12. One point is assigned for each 0.25 percent of building energy cost savings. The credits can be accumulated across a total of 11 options.

In addition to the significance of the revised structure of the section, the following options are either new or significantly revised:

The previous HVAC additional efficiency measure of C406.2 was modified to include several credit options that apply to heating equipment, cooling equipment or both, with separate credits based on improvement in efficiency.

- Medium efficiency HVAC heating performance (C406.2.1) is a 5 percent improvement in efficiency over the existing minimum requirement.
- Medium efficiency HVAC cooling performance (C406.2.2) is a 5 percent improvement in efficiency over the existing minimum requirement.

- High efficiency HVAC heating performance (C406.2.3) is a 10 percent improvement in efficiency over the existing minimum requirement.
- High efficiency HVAC cooling performance (C406.2.4) is a 10 percent improvement in efficiency over the existing minimum requirement.
- High efficiency HVAC cooling performance (C406.2.5) is more than a 10 percent improvement in efficiency over the existing minimum requirement.

The lighting power reduction option in C406.3 has been expanded to allow for additional credits as well. Previously, there was a single option to reduce lighting power density by 10 percent; the 2021 IECC now provides options for reducing lighting power by 15 percent and installing 95 percent high efficacy lighting in dwelling and sleeping units.

The on-site renewable energy option in Section C406.5 has been modified to include two levels of credits for renewable energy. The basic credit is achieved with .25 watts/ft^2 of conditioned floor area or at least 2 percent of the mechanical, service water heating and lighting annual energy. Enhanced credit is achieved with increased on-site renewable energy systems.

Code users should note the correlated revised definition:

> **ON-SITE RENEWABLE ENERGY.** Energy derived from renewable energy resources harvested at the building project site.

The 2018 IECC language provided for energy recovery in Section C406.7 has been revised in the 2021 IECC to provide credit for high efficiency gas and electric water heaters in addition to heat recovery. These credits are available to seven building occupancy types as well as buildings for which service hot water heating accounts for 10 or more percent of the total building energy loads.

The 2021 IECC includes energy monitoring requirements in Section C405.12. The requirement does not apply to buildings less than 25,000 square feet, and R-2 occupancies and individual tenant spaces are exempt if the space has its own utility services and meters and has less than 5,000 square feet of conditioned floor area. Section C406.10 provides additional efficiency credit options for buildings not otherwise required to have monitoring by Section C405.12.

The energy efficiency of a new building's HVAC system will degrade over time when equipment is poorly maintained, failing or improperly controlled. The new fault detection and diagnostics (FDD) requirement in C406.11 will reduce that degradation by detecting HVAC system faults and notifying building operators so that actions may be taken to reduce the energy consumption of the building. Additionally, FDD systems are being utilized to drive operational efficiency, make better use of maintenance personnel and resolve comfort issues.

The final option of the revised Section C406 is the installation of efficient kitchen equipment in Group A-2 or facilities that have a commercial kitchen with at least one gas or electric fryer (C406.12). Kitchen equipment accounts for a large share of building energy use in restaurants, schools, dormitories, hotels and other facilities with full-service kitchens. More efficient equipment saves energy by improving the heat transfer in the cooking process, either through better equipment insulation or other innovations in the appliances. This option is different from the other 10 in that the value of the credit is not determined using Tables C406.1(1) through C406.1(5) and instead is calculated using Equation 4-15.

The additional efficiency credit approach retains designer and builder flexibility, and using points more closely aligns the various credit options across climate zones. In some cases, however, the options also require additional documentation and review by code officials.

Section C406 and the new and revised options included within are applicable only to the Prescriptive Compliance path.

C407

Total Building Performance

The Total Building Performance provisions of Section C407 comprise the second energy code compliance option of the IECC. This compliance method uses computer software to predict and compare the annual energy cost of the proposed design to the standard design of the building using the components listed in Table C407.4.1(1). Total Building Performance allows tradeoffs between the building envelope, HVAC systems and lighting systems to achieve an overall building energy conservation result equal to or better than that of the other code compliance paths.

The significant change to the Total Building Performance option is the addition of a new, comprehensive table of required provisions.

C407.2

Performance-Based Compliance

CHANGE TYPE: Modification

CHANGE SUMMARY: Code requirements applicable to the Total Building Performance compliance path are outlined in a new, user-friendly table to reduce code application confusion. The proposed design for projects complying with Total Building Performance must have an annual energy cost less than or equal to 85 percent of the standard reference design.

2021 CODE: **C407.2 Mandatory requirements.** ~~Compliance with this section requires compliance with Sections C402.2.7, C402.5, C403.2, C403.3 through C403.3.2, C403.4 through C403.4.2.3, C403.5.5, C403.7, C403.8.1 through C403.8.4, C403.10.1 through C403.10.3, C403.11, C403.12, C404 and C405~~ ~~C405.1, C405.2, C405.4 through C405.9, and C408~~. Compliance based on total building performance requires that a proposed design meet all of the following: ~~building (proposed design) be shown to have~~

1. The requirements of the sections indicated within Table C407.2.
2. An annual energy cost that is less than or equal to 85 percent of the annual energy cost of the standard reference design. Energy prices shall be taken from a source approved by the code official, such as the Department of Energy, Energy Information Administration's State Energy ~~Price and Expenditure Report~~ Data System Prices and Expenditures reports. Code officials shall be permitted to require time-of-use pricing in energy cost calculations. The reduction in energy cost of the proposed design associated with on-site renewable energy shall be not more than 5 percent of the total energy cost. The amount of renewable energy purchased from off-site sources shall be the same in the standard reference design and the proposed design.

Exception: Jurisdictions that require site energy (1 kWh = 3413 Btu) rather than energy cost as the metric of comparison.

Photo courtesy of APCortizasJr/E+/Getty Images

Requirements for Total Building Performance compliance are clarified.

TABLE C407.2 Requirements for Total Building Performance

SECTION[a]	TITLE
	Envelope
C402.5	Air leakage—thermal envelope
	Mechanical
C403.1.1	Calculation of heating and cooling loads
C403.1.2	Data centers
C403.2	System design
C403.3	Heating and cooling equipment efficiencies
C403.4, except C403.4.3, C403.4.4 and C403.4.5	Heating and cooling system controls
C403.5.5	Economizer fault detection and diagnostics
C403.7, except C403.7.4.1	Ventilation and exhaust systems
C403.8, except C403.8.6	Fan and fan controls
C403.9	Large-diameter ceiling fans
C403.11, except C403.11.3	Refrigeration equipment performance
C403.12	Construction of HVAC system elements
C403.13	Mechanical systems located outside of the building thermal envelope
C404	Service water heating
C405, except C405.3	Electrical power and lighting systems
C408	Maintenance information and system commissioning

a. Reference to a code section includes all the relative subsections except as indicated in the table.

CHANGE SIGNIFICANCE: The labels *prescriptive* and *mandatory* were introduced in the 2009 IECC and, despite never being defined, were applied to various section and subsection titles throughout the IECC, creating confusion for users of the code. The terms were used inconsistently and because the labels themselves were not requirements, the intent behind the designation was not enforceable.

Among long-time code users, it is generally understood that *mandatory* was intended to mean "non-tradeable" when using performance compliance options; that is, where the procedures or systems described within the 'mandatory' section were included as part of the design, the requirements of that section must be met and cannot be traded off. *Prescriptive*, on the other hand, was intended to mean required when using the prescriptive path but "tradeable" when using the performance path. However, this understanding was not explicitly based on code language, and new users to the code expressed confusion about the use of the terms.

The 2021 IECC eliminates the terms mandatory and prescriptive, and instead, all requirements applicable to the Total Building Performance compliance path are now located in Section C407. Table C407.2 includes all required provisions for this compliance path.

This provision is applicable to the Total Building Performance compliance option only.

Calvin Chan Wai Meng/ Moment/Getty Images

PART 4

Commercial Energy Provisions: Commercial Appendices

- **Appendix CA** Board of Appeals – Commercial
- **Appendix CB** Solar-Ready Zone – Commercial
- **Appendix CC** Zero Energy Commercial Building Provisions

Appendices to the *International Energy Conservation Code* (IECC) provide options for jurisdictions to expand the requirements of the IECC. These options are not mandatory unless specifically referenced in the adopting ordinances. The 2021 IECC introduces two new commercial appendices - Appendix CA provides guidance for establishing a board of appeals and Appendix CC outlines provisions for zero energy commercial buildings. The existing solar-ready appendix is also modified. ■

Appendix CA

Board of Appeals – Commercial

CB103.7

Electrical Energy Storage System-Ready Area

Appendix CC

Zero Energy Commercial Building Provisions

Appendix CA

Board of Appeals – Commercial

CHANGE TYPE: Addition

CHANGE SUMMARY: A new Appendix CA provides guidance for establishing a board of appeals, including criteria for membership and instruction for developing rules and procedures.

2021 CODE:

APPENDIX CA
BOARD OF APPEALS—COMMERCIAL

SECTION CA101
GENERAL

CA101.1 Scope. A board of appeals shall be established within the jurisdiction for the purpose of hearing applications for modification of the requirements of this code pursuant to the provisions of Section C110. The board shall be established and operated in accordance with this section, and shall be authorized to hear evidence from appellants and the code official pertaining to the application and intent of this code for the purpose of issuing orders pursuant to these provisions.

CA101.2 Application for appeal. Any person shall have the right to appeal a decision of the code official to the board. An application for appeal shall be based on a claim that the intent of this code or the rules legally adopted hereunder have been incorrectly interpreted, the provisions of this code do not fully apply or an equally good or better form of construction is proposed. The application shall be filed on a form obtained from the code official within 20 days after the notice was served.

CA101.2.1 Limitation of authority. The board shall not have authority to waive requirements of this code or interpret the administration of this code.

Photo courtesy of AzmanL

A board of appeals reviews evidence from appellants and code officials.

CA101.2.2 Stays of enforcement. Appeals of notice and orders, other than Imminent Danger notices, shall stay the enforcement of the notice and order until the appeal is heard by the board.

CA101.3 Membership of board. The board shall consist of five voting members appointed by the chief appointing authority of the jurisdiction. Each member shall serve for [INSERT NUMBER OF YEARS] years or until a successor has been appointed. The board members' terms shall be staggered at intervals, so as to provide continuity. The code official shall be an ex officio member of said board but shall not vote on any matter before the board.

CA101.3.1 Qualifications. The board shall consist of five individuals, who are qualified by experience and training to pass on matters pertaining to building construction and are not employees of the jurisdiction.

CA101.3.2 Alternate members. The chief appointing authority is authorized to appoint two alternate members who shall be called by the board chairperson to hear appeals during the absence or disqualification of a member. Alternate members shall possess the qualifications required for board membership, and shall be appointed for the same term or until a successor has been appointed.

CA101.3.3 Vacancies. Vacancies shall be filled for an unexpired term in the same manner in which original appointments are required to be made.

CA101.3.4 Chairperson. The board shall annually select one of its members to serve as chairperson.

CA101.3.5 Secretary. The chief appointing authority shall designate a qualified clerk to serve as secretary to the board. The secretary shall file a detailed record of all proceedings which shall set forth the reasons for the board's decision, the vote of each member, the absence of a member and any failure of a member to vote.

CA101.3.6 Conflict of interest. A member with any personal, professional or financial interest in a matter before the board shall declare such interest and refrain from participating in discussions, deliberations and voting on such matters.

CA101.3.7 Compensation of members. Compensation of members shall be determined by law.

CA101.3.8 Removal from the board. A member shall be removed from the board prior to the end of their terms only for cause. Any member with continued absence from regular meeting of the board may be removed at the discretion of the chief appointing authority.

CA101.4 Rules and procedures. The board shall establish policies and procedures necessary to carry out its duties consistent with the provisions of this code and applicable state law. The procedures shall not require compliance with strict rules of evidence, but shall mandate that only relevant information be presented.

CA101.5 Notice of meeting. The board shall meet upon notice from the chairperson, within 10 days of the filing of an appeal or at stated periodic intervals.

CA101.5.1 Open hearing. All hearings before the board shall be open to the public. The appellant, the appellant's representative, the code official and any person whose interests are affected shall be given an opportunity to be heard.

CA101.5.2 Quorum. Three members of the board shall constitute a quorum.

CA101.5.3 Postponed hearing. When five members are not present to hear an appeal, either the appellant or the appellant's representative shall have the right to request a postponement of the hearing.

CA101.6 Legal counsel. The jurisdiction shall furnish legal counsel to the board to provide members with general legal advice concerning matters before them for consideration. Members shall be represented by legal counsel at the jurisdiction's expense in all matters arising from service within the scope of their duties.

CA101.7 Board decision. The board shall only modify or reverse the decision of the code official by a concurring vote of three or more members.

CA101.7.1 Resolution. The decision of the board shall be by resolution. Every decision shall be promptly filed in writing in the office of the code official within three days and shall be open to the public for inspection. A certified copy shall be furnished to the appellant or the appellant's representative and to the code official.

CA101.7.2 Administration. The code official shall take immediate action in accordance with the decision of the board.

CA101.8 Court review. Any person, whether or not a previous party of the appeal, shall have the right to apply to the appropriate court for a writ of certiorari to correct errors of law. Application for review shall be made in the manner and time required by law following the filing of the decision in the office of the chief administrative officer.

CHANGE SIGNIFICANCE: Appendix CA mirrors the requirements of the 2018 *International Building Code* (IBC), 2018 *International Fire Code* (IFC) and 2021 *International Residential Code* (IRC), providing a consistent set of requirements for a board of appeals across the family of I-Codes. This appendix offers a framework for jurisdictions to establish a functional board of appeals, outlining the rights and responsibilities of the appellant, code official and board members. The provisions contained in this appendix are not mandatory unless specified as such in the jurisdiction's adopting ordinance.

CB103.7

Electrical Energy Storage System-Ready Area

CHANGE TYPE: Addition

CHANGE SUMMARY: Appendix CB adds a provision for the design and construction of a system-ready area for electrical energy storage.

2021 CODE: **CB103.7 Electrical energy storage system-ready area.** The floor area of the electrical energy storage system-ready area shall be not less than 2 feet (610 mm) in one dimension and 4 feet (1219 mm) in another dimension, and located in accordance with Section 1207 of the International Fire Code. The location and layout diagram of the electrical energy storage system-ready area shall be indicated on the construction documents.

CHANGE SIGNIFICANCE: The Solar-Ready Zone appendices to the IECC are useful for jurisdictions seeking to add solar-ready provisions to state or local codes. These appendices provide requirements for reserving sections of the roof or building overhang for future solar photovoltaics or solar thermal systems. Many jurisdictions considering the Solar-Ready Zone appendices are facing current or future constraints on electric grid capacity in their efforts to accommodate existing and new distributed solar generation resources; as such, policy objectives are emerging to support the storage of energy produced by solar panels and shift the temporal impact of such energy on the grid.

The addition of Section CB103.7 ensures that there are design and space considerations for a standard-sized battery rack and for connections to the electrical panels. As with the rationale for solar-readiness, it is generally much more cost-effective at the time of new construction to design for future installation of this equipment than it is to retrofit later.

Photo courtesy of Vithun Khamsong/Moment/Getty Images

Buildings in jurisdictions adopting this appendix must be energy storage ready.

Section CB103.7 also cites the *International Fire Code* (IFC) to ensure there is sufficient clearance around the battery rack to meet life/safety concerns. The IFC is also included in Chapter 6 as a referenced code. The provisions contained in this appendix are not mandatory unless specified as such in the jurisdiction's adopting ordinance.

Appendix CC

Zero Energy Commercial Building Provisions

CHANGE TYPE: Addition

CHANGE SUMMARY: Appendix CC provides a model for states and jurisdictions to require renewable energy systems capable of achieving net zero carbon.

2021 CODE:

APPENDIX CC
ZERO ENERGY COMMERCIAL BUILDING PROVISIONS

SECTION CC101
GENERAL

CC101.1 Purpose. The purpose of this appendix is to supplement the International Energy Conservation Code and require renewable energy systems of adequate capacity to achieve net zero carbon.

CC101.2 Scope. This appendix applies to new buildings that are addressed by the International Energy Conservation Code.

Exceptions:

1. Detached one- and two-family dwellings and townhouses as well as Group R-2 buildings three stories or less in height above grade plane, manufactured homes (mobile dwellings), and manufactured houses (modular dwellings).
2. Buildings that use neither electricity nor fossil fuel.

SECTION CC102
DEFINITIONS

CC102.1 Definitions. The definitions contained in this section supplement or modify the definitions in the International Energy Conservation Code.

Photo courtesy of KenWiedemann/E+/Getty Images

On-site renewables.

ADJUSTED OFF-SITE RENEWABLE ENERGY. The amount of energy production from off-site renewable energy systems that may be used to offset building energy.

BUILDING ENERGY. All energy consumed at the building site as measured at the site boundary. Contributions from on-site or off-site renewable energy systems shall not be considered when determining the building energy.

ENERGY UTILIZATION INTENSITY (EUI). The site energy for either the baseline building or the proposed building divided by the gross conditioned floor area plus any semiheated floor area of the building. For the baseline building, the EUI can be divided between regulated energy use and unregulated energy use.

OFF-SITE RENEWABLE ENERGY SYSTEM. Renewable energy system not located on the building project.

ON-SITE RENEWABLE ENERGY SYSTEM. Renewable energy systems on the building project.

RENEWABLE ENERGY SYSTEM. Photovoltaic, solar thermal, geothermal energy and wind systems used to generate energy.

SEMIHEATED SPACE. An enclosed space within a building that is heated by a heating system whose output capacity is greater than or equal to 3.4 Btu/h×ft^2 of floor area but is not a conditioned space.

ZERO ENERGY PERFORMANCE INDEX (ZEPI PB/EE). The ratio of the proposed building EUI without renewables to the baseline building EUI, expressed as a percentage.

SECTION CC 103
MINIMUM RENEWABLE ENERGY

CC103.1 Renewable Energy. On-site renewable energy systems shall be installed, or off-site renewable energy shall be procured to offset the building energy as calculated in accordance with Equation CC-1.

$$RE_{onsite} + RE_{offsite} \geq E_{building} \qquad \textbf{(Equation CC-1)}$$

where:

RE_{onsite} = Annual site energy production from on-site renewable energy systems (see Section CC103.2).

$RE_{offsite}$ = Adjusted annual site energy production from off-site renewable energy systems that may be credited against building energy use (see Section CC103.3).

$E_{building}$ = Building energy use without consideration of renewable energy systems.

When Section C401.2.1(1) is used for compliance with the International Energy Conservation Code, building energy shall be determined by multiplying the gross conditioned floor area plus the gross semiheated floor area of the proposed building by an EUI selected from Table CC103.1. Use a weighted average for mixed-use buildings.

When Section C401.2.1, Item 2 or Section C401.2.2 is used for compliance with the International Energy Conservation Code, building energy shall be determined from energy simulations.

TABLE CC103.1 Energy Utilization Intensity for Building Types and Climates (kBtu/ft² – yr)

Building Area Type	Climate Zone																
	0A/ 1A	0B/ 1B	2A	2B	3A	3B	3C	4A	4B	4C	5A	5B	5C	6A	6B	7	8
	kBtu/ft²-yr																
Healthcare/hospital (I-2)	119	120	119	113	116	109	106	116	109	106	118	110	105	126	116	131	142
Hotel/motel (R-1)	73	76	73	68	70	67	65	69	66	65	71	68	65	77	72	81	89
Multiple-family (R-2)	43	45	41	41	43	42	36	45	43	41	47	46	41	53	48	53	59
Office (B)	31	32	30	29	29	28	25	28	27	25	29	28	25	33	30	32	36
Restaurant (A-2)	389	426	411	408	444	420	395	483	437	457	531	484	484	589	538	644	750
Retail (M)	46	50	45	46	44	44	37	48	44	44	52	50	46	60	52	64	77
School (E)	42	46	42	40	40	39	36	39	40	40	39	43	37	44	40	45	54
Warehouse (S)	9	12	9	11	12	11	10	17	13	14	23	17	15	32	23	32	32
All others	55	58	54	53	53	51	48	54	52	51	57	54	50	63	57	65	73

CC103.2 Calculation of on-site renewable energy. The annual energy production from on-site renewable energy systems shall be determined using the PVWatts software or other software approved by the code official.

CC103.3 Off-Site renewable energy. Off-site energy shall comply with Sections CC103.3.1 and CC103.3.2.

CC103.3.1 Qualifying off-site procurement methods. The following are considered qualifying off-site renewable energy procurement methods:

1. Community renewables: an offsite renewable energy system for which the owner has purchased or leased renewable energy capacity along with other subscribers.
2. Renewable energy investment fund: an entity that installs renewable energy capacity on behalf of the owner.
3. Virtual power purchase agreement: a power purchase agreement for off-site renewable energy where the owner agrees to purchase renewable energy output at a fixed price schedule.
4. Direct ownership: an offsite renewable energy system owned by the building project owner.
5. Direct access to wholesale market: an agreement between the owner and a renewable energy developer to purchase renewable energy.
6. Green retail tariffs: a program by the retail electricity provider to provide 100-percent renewable energy to the owner.

7. Unbundled Renewable Energy Certificates (RECs): certificates purchased by the owner representing the environmental benefits of renewable energy generation that are sold separately from the electric power.

CC103.3.2 Requirements for all procurement methods. The following requirements shall apply to all off-site renewable energy procurement methods:

1. The building owner shall sign a legally binding contract to procure qualifying off-site renewable energy.
2. The procurement contract shall have duration of not less than 15 years and shall be structured to survive a partial or full transfer of ownership of the property.
3. RECs and other environmental attributes associated with the procured off-site renewable energy shall be assigned to the building project for the duration of the contract.
4. The renewable energy generating source shall include one or more of the following: photovoltaic systems, solar thermal power plants, geothermal power plants and wind turbines.
5. The generation source shall be located where the energy can be delivered to the building site by the same utility or distribution entity, the same independent system operator (ISO) or regional transmission organization (RTO), or within integrated ISOs (electric coordination council).
6. The off-site renewable energy producer shall maintain transparent accounting that clearly assigns production to the building. Records on power sent to or purchased by the building shall be retained by the building owner and made available for inspection by the code official upon request.

CC103.3.3 Adjusted off-site renewable energy. The process for calculating the adjusted off-site renewable energy is shown in the Equation CC-2.

$$RE_{offsite} = \sum_{i=1}^{n} PF_i \cdot RE_i = PF_1 \cdot RE_1 + PF_2 \cdot RE_2 + \cdots + PF_n \cdot RE_n$$

(Equation CC-2)

where:

$RE_{offsite}$ = Adjusted off-site renewable energy.

PF_i = Procurement factor for the ith renewable energy procurement method or class taken from Table CC103.3.3.

RE_i = Annual energy production for the ith renewable energy procurement method or class.

n = The number of renewable energy procurement options or classes considered.

TABLE CC103.3.3 Default Off-Site Renewable Energy Procurement Methods, Classes and Coefficients

Class	Procurement Factor (PF)	Procurement Options	Additional Requirements (see also Section CC103.3.2)
1	0.75	Community Solar	—
		REIFs	Entity must be managed to prevent fraud or misuse of funds.
		Virtual PPA	—
		Self-Owned Off-Site	Provisions shall prevent the generation from being sold separately from the building.
2	0.55	Green Retail Tariffs	The offering shall not include the purchase of unbundled RECs.
		Direct Access	The offering shall not include the purchase of unbundled RECs.
3	0.20	Unbundled RECs	The vintage of the RECs shall align with building energy use.

CHANGE SIGNIFICANCE: The relationship between renewable energy systems and code compliance has changed throughout the years. The 2003 IECC allowed full attribution of on-site renewable energy in the performance compliance path. More recently, however, the amount of renewable energy attributable to compliance has been limited. For example, the 2018 IECC includes the use of renewable energy as an additional efficiency package in C406, and on-site renewable energy is limited to 5 percent of the total energy cost in C407 Total Building Performance. Similarly, additional backstops compensate for the use of on-site renewables in the R406 Energy Rating Index compliance alternative for residential projects.

Appendix CC is based on Architecture 2030, which focuses on carbon emissions reduction and couples onsite and/or off-site renewable energy with energy efficiency measures resulting in Zero-Net-Carbon (ZNC) buildings. Architecture 2030 defines a ZNC building as, "a highly energy efficient building that produces on-site, or procures, enough carbon-free renewable energy to meet building operations energy consumption annually." Like Architecture 2030, Appendix CC requires compliance with the minimum provisions of the IECC and then provides calculations to determine the minimum renewable energy necessary to offset building energy use.

As such, Appendix CC once again fully recognizes the input of renewable energy for commercial projects; however, it does not expand the attribution of renewable energy in minimum code compliance.

Appendix CC provides jurisdictions an option for establishing net-zero carbon targets. The provisions contained in this appendix are not mandatory unless specified as such in the jurisdiction's adopting ordinance.

PART 5

Residential Energy Provisions: Administration

Chapters 1 and 2

Photo courtesy of artursfoto/iStock / Getty Images Plus/ Getty Images

The administration section of the *International Energy Conservation Code* (IECC) covers the general scope, intent, applicability and enforcement of the IECC. Section R101.2 establishes the criteria for buildings that are regulated by the IECC Residential provisions. Buildings outside the scope of Section R101.2 are regulated by the Commercial provisions of the IECC. The administration provisions of Chapter 1 cover the applicability of the code to mixed-use buildings; alternative materials; design and methods of construction; construction documents; inspection fees; referenced standards and stop work orders.

Definitions used throughout the IECC are found in Chapter 2. The definitions contained within the IECC are intended to reflect the special meaning of such terms within the scope of the code. As terms can often have multiple meanings in their ordinary day-to-day use or within the various disciplines of the construction industry, it is important that their meanings be understood within the context of the IECC. ■

R103.1

Digital Plans

R103.2

Information on Construction Documents

R202

Definition of High-Efficacy Light Sources

R202

Definition of Renewable Energy Certificate

R103.1

Digital Plans

CHANGE TYPE: Clarification

CHANGE SUMMARY: Provisions related to energy code documentation clarify that construction documents and supporting data may be submitted digitally.

2021 CODE: R103.1 General. Construction documents, technical reports and other supporting data shall be submitted in one or more sets, or in a digital format where allowed by the code official, with each application for a permit. The construction documents shall be prepared by a registered design professional where required by the statutes of the jurisdiction in which the project is to be constructed. Where special conditions exist, the code official is authorized to require necessary construction documents to be prepared by a registered design professional.

> **Exception:** The code official is authorized to waive the requirements for construction documents or other supporting data if the code official determines they are not necessary to confirm compliance with this code.

CHANGE SIGNIFICANCE: This provision was first drafted when blueprints were the standard, and construction plans and documentation were submitted on paper. As more jurisdictions move to an online, digital submittal process, the construction documents provisions proved confusing because the requirement for "one or more sets" implied to some that paper copies are still required. This code change clarifies that the code official may choose to accept digital submittals in lieu of paper documentation. The digital format facilitates a comprehensive electronic permitting, plan review and inspection system, and includes these benefits:

- Streamlined online submittals, payments, reviews and inspection scheduling
- Standardized building site information
- Readily available project status and information
- Real-time inspection results from the field.

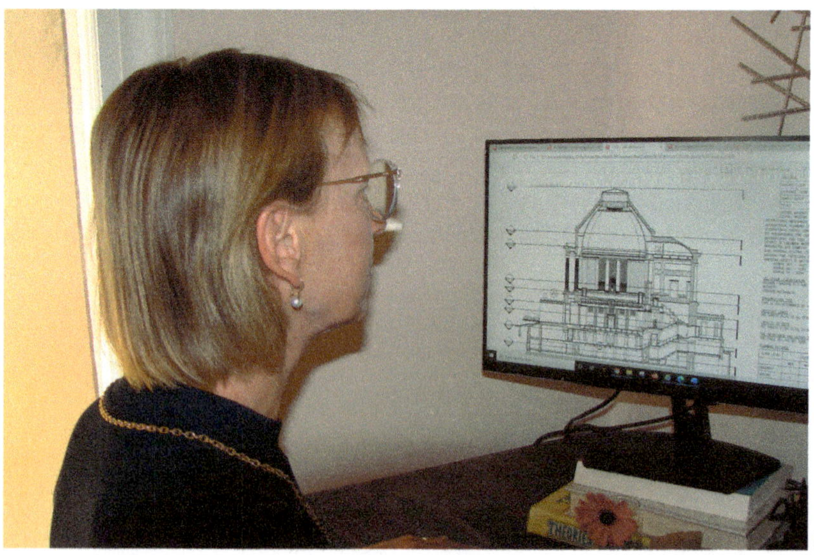

Photo courtesy of Wendy Talarico

Digital plan review.

R103.2

Information on Construction Documents

CHANGE TYPE: Modification

CHANGE SUMMARY: The information required on the construction documents has been expanded to include an indication of the energy compliance path used.

2021 CODE: **R103.2 Information on construction documents.** Construction documents shall be drawn to scale on suitable material. Electronic media documents are permitted to be submitted where approved by the code official. Construction documents shall be of sufficient clarity to indicate the location, nature and extent of the work proposed, and show in sufficient detail pertinent data and features of the building, systems and equipment as herein governed. Details shall include the following as applicable:

1. Energy compliance path.
2. Insulation materials and their *R*-values.
3. Fenestration *U*-factors and solar heat gain coefficients (SHGCs).
4. Area-weighted *U*-factor and solar heat gain coefficient (SHGC) calculations.
5. Mechanical system design criteria.
6. Mechanical and service water-heating systems and equipment types, sizes and efficiencies.
7. Equipment and system controls.
8. Duct sealing, duct and pipe insulation and location.
9. Air sealing details.

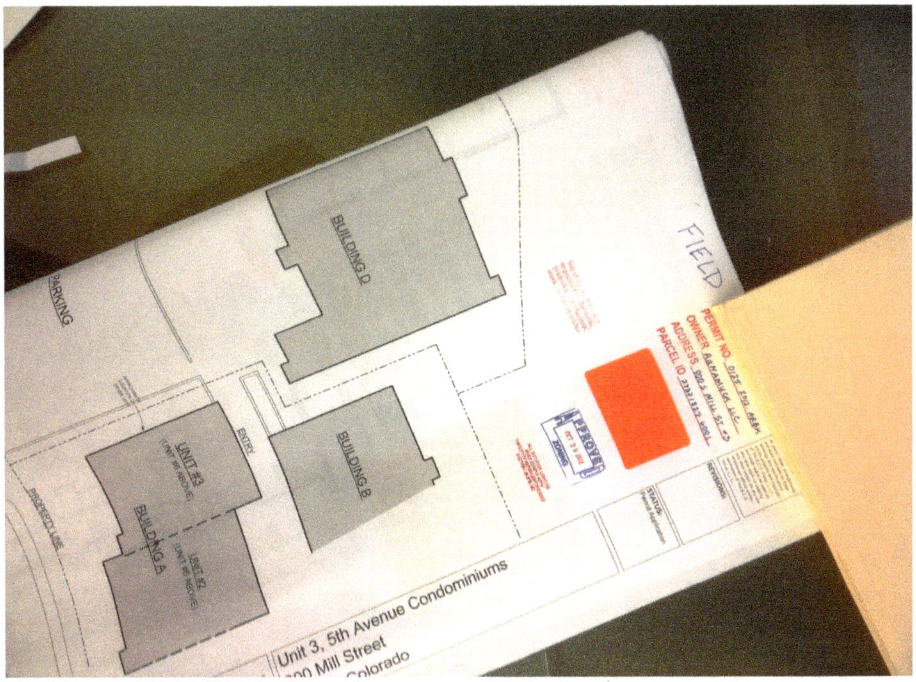

Energy compliance path must be noted on construction documents.

CHANGE SIGNIFICANCE: Identification of the compliance path is essential for the plan reviewer to determine if the project demonstrates compliance with the specific energy requirements. When this information is not provided on the construction documents and plans examiners are required to investigate, the review process is lengthened. This extension may be in the form of a review comment or the time it takes to retrieve the information by email or phone. Providing this information at submittal allows the plans examiner to review the plans according to the intended energy compliance path selected by the architect or designer.

R202

Definition of High-Efficacy Light Sources

CHANGE TYPE: Modification

CHANGE SUMMARY: The revised definition of high-efficacy light sources increases efficacy to 65 lumens per watt for lamps and 45 lumens per watt for luminaires.

2021 CODE: ~~**HIGH-EFFICACY LAMPS.**~~ **HIGH-EFFICACY LIGHT SOURCES.** ~~Compact fluorescent lamps, light-emitting diode (LED) lamps, T-8 or smaller diameter linear fluorescent lamps, or other lamps~~ Any lamp with an efficacy of not less than ~~the following:~~ 65 lumens per watt, or luminaires with an efficacy of not less than 45 lumens per watt.

~~1. 60 lumens per watt for lamps over 40 watts~~

~~2. 50 lumens per watt for lamps over 15 watts to 40 watts.~~

~~3. 40 lumens per watt for lamps 15 watts or less.~~

CHANGE SIGNIFICANCE: High-efficacy lamps were first defined in the 2009 IECC when the LED market was in its infancy. Since then, lighting technologies have evolved and lighting standards have changed such that many of the technologies defined by the 2009 IECC are no longer compliant. The 2021 IECC updates the definition of high-efficacy light sources to capture these market changes and to provide separate thresholds for lamps and luminaires.

Photo courtesy of Peter Dazeley/The Image Bank/Getty Images

High-efficacy light sources are required in all permanently installed lighting fixtures.

R202

Definition of Renewable Energy Certificate

CHANGE TYPE: Addition

CHANGE SUMMARY: A new definition for Renewable Energy Certificate ensures homeowners can provide legal documentation of their ownership of the credits for renewable energy.

2021 CODE: **RENEWABLE ENERGY CERTIFICATE (REC).** An instrument that represents the environmental attributes of one megawatt hour of renewable energy; also known as an energy attribute certificate (EAC).

CHANGE SIGNIFICANCE: Renewable Energy Certificates (RECs) represent the environmental attributes of renewable energy. As the on-site renewable market continues to evolve, RECs capture the current market value of solar power and other renewable energy used for residential projects.

The 2021 IECC requires homeowners using on-site renewable energy for compliance with the Energy Rating Index to own the RECs for the home. This requirement ensures the environmental attributes of the RECs are counted toward one home only. According to the Federal Trade Commission, when the installer, leasing company, financial agent or other entity in the renewable energy transaction owns the REC, the power produced by the system does not qualify as renewable energy. This is because the REC may be sold to more than one party, resulting in a double-counting of the environmental attributes of the system.

Photo courtesy of cdwheatley/iStock / Getty Images Plus/ Getty Images

On-site renewable energy.

PART 6

Residential Energy Provisions: General Requirements

Chapter 3

Photo courtesy of BanksPhotos/E+/Getty Images

- **Chapter 3** General Requirements

R301.3

Climate Zone Definitions

Chapter 3 provides foundational requirements for provisions in Chapters 4 and 5.

Climate zone designations in Chapter 3 are used as a reference for the building envelope and mechanical provisions that are climate-specific. The code establishes many requirements, such as wall and roof insulation *R*-values and window and door thermal transmittance (*U*-factors), that affect the mechanical systems based on the climate where the building is located.

In addition, Chapter 3 provides interior design conditions that are used as a basis for assumptions in heating and cooling load calculations and basic material requirements for insulation and fenestration. Requirements for labeling and product rating are also provided; these are foundational to the envelope requirements of Chapter 4. Finally, Chapter 3 provides specifications for insulation installation that help ensure the effectiveness of the insulation. ■

R301.3

Climate Zone Definitions

CHANGE TYPE: Modification

CHANGE SUMMARY: The climate zone provisions have been comprehensively updated and include a new Climate Zone 0. Ten percent of U.S. counties are now in a new climate zone.

2021 CODE TEXT: **R301.3** ~~**International climate zones**~~ **Climate zone definitions.** ~~The climate zone for any location outside the United States shall be determined by applying Table R301.3(1) and then Table R301.3(2).~~ To determine the climate zones for locations not listed in this code, use the following information to determine climate zone numbers and letters in accordance with Items 1 through 5.

1. Determine the thermal climate zone, 0 through 8, from Table R301.3 using the heating (HDD) and cooling degree-days (CDD) for the location.
2. Determine the moisture zone (Marine, Dry or Humid) in accordance with Items 2.1 through 2.3.
   - 2.1. If monthly average temperature and precipitation data are available, use the Marine, Dry, and Humid definitions to determine the moisture zone (C, B or A).
   - 2.2. If annual average temperature information (including degree-days) and annual precipitation (i.e., annual mean) are available, use Items 2.2.1 through 2.2.3 to determine the moisture zone. If the moisture zone is not Marine, then use the Dry definition below to determine whether Dry or Humid:
     - 2.2.1. If thermal climate zone is 3 and CDD50°F ≤ 4,500 (CDD10°C ≤2500), climate zone is Marine (3C).
     - 2.2.2. If thermal climate zone is 4 and CDD50°F ≤ 2,700 (CDD10°C ≤ 1500), climate zone is Marine (4C).
     - 2.2.3. If thermal climate zone is 5 and CDD50°F ≤ 1,800 (CDD10°C ≤ 1000), climate zone is Marine (5C).

The IECC provisions are climate specific.

2.3. If only degree-day information is available, use Items 2.3.1 through 2.3.3 to determine the moisture zone. If the moisture zone is not Marine, then it is not possible to assign Humid or Dry moisture zone for this location.

2.3.1. If thermal climate zone is 3 and CDD50°F ≤ 4,500 (CDD10°C ≤ 2500), climate zone is Marine (3C).

2.3.2. If thermal climate zone is 4 and CDD50°F ≤ 2,700 (CDD10°C ≤ 1500), climate zone is Marine (4C).

2.3.3. If thermal climate zone is 5 and CDD50°F ≤ 1,800 (CDD10°C ≤ 1000), climate zone is Marine (5C).

3. Marine (C) Zone definition: Locations meeting all the criteria in Items 3.1 through 3.4.

3.1. Mean temperature of coldest month between 27°F (−3°C) and 65°F (18°C).

3.2. Warmest month mean < 72°F (22°C).

3.3. Not fewer than four months with mean temperatures over 50°F (10°C).

3.4. Dry season in summer. The month with the heaviest precipitation in the cold season has at least three times as much precipitation as the month with the least precipitation in the rest of the year. The cold season is October through March in the Northern Hemisphere and April through September in the Southern Hemisphere.

4. Dry (B) definition: Locations meeting the criteria in Items 4.1 through 4.4.

4.1. Not Marine (C).

4.2. If 70 percent or more of the precipitation, P, occurs during the high sun period, defined as April through September in the Northern Hemisphere and October through March in the Southern Hemisphere, then the dry/humid threshold is in accordance with Equation 3-1.

$P < 0.44 \times (T - 7)$

$[P < 20.0 \times (T + 14)$ in SI units$]$ (Equation 3-1)

where:

P = Annual precipitation, inches (mm).
T = Annual mean temperature, °F (°C).

4.3. If between 30 and 70 percent of the precipitation, P, occurs during the high sun period, defined as April through September in the Northern Hemisphere and October through March in the Southern Hemisphere, then the dry/humid threshold is in accordance with Equation 3-2.

$P < 0.44 \times (T - 19.5)$

$[P < 20.0 \times (T + 7)$ in SI units$]$ (Equation 3-2)

where:

P = Annual precipitation, inches (mm).

T = Annual mean temperature, °F (°C).

4.4. If 30 percent or less of the precipitation, P, occurs during the high sun period, defined as April through September in the Northern Hemisphere and October through March in the Southern Hemisphere, then the dry/humid threshold is in accordance with Equation 3-3.

$P < 0.44 \times (T - 32)$

$[P < 20 \times T$ in SI units$]$ (Equation 3-3)

where:

P = Annual precipitation, inches (mm).

T = Annual mean temperature, °F (°C).

5. Humid (A) definition: Locations that are not Marine (C) or Dry (B).

TABLE R301.1 Climate Zones, Moisture Regimes, and Warm Humid Designations by State, County and Territory[a]

US STATES

ALABAMA
~~3A~~ 2A Coffee*
~~3A~~ 2A Covington*
~~3A~~ 2A Dale*
~~3A~~ 2A Escambia*
~~3A~~ 2A Geneva*
~~3A~~ 2A Henry*
~~3A~~ 2A Houston*

ALASKA
~~8~~ 7 Bethel
~~7~~ 8 Denali
~~8~~ 7 Dillingham
~~7~~ 6A Haines
~~7~~ 6A Juneau
~~7~~ 5C Ketchikan Gateway
~~7~~ 6A Kodiak Island
~~7~~ 5C Prince of Wales Outer Ketchikan
~~7~~ 5C Sitka
~~7~~ 6A Skagway-Hoonah-Angoon
~~7~~ 6A Wrangell-Petersburg

ARIZONA

ARKANSAS

CALIFORNIA

COLORADO
~~5B~~ 4B Bent
~~6B~~ 5B Custer
~~5B~~ 4B Prowers

CONNECTICUT

DELAWARE

DISTRICT OF COLUMBIA

FLORIDA
~~2A~~ 1A Palm Beach*

GEORGIA
~~4A~~ 3A Banks
~~3A~~ 2A Calhoun*
~~4A~~ 3A Catoosa
~~4A~~ 3A Chattooga
~~3A~~ 2A Coffee*
~~4A~~ 3A Dade
~~4A~~ 3A Dawson
~~3A~~ 2A Dougherty*
~~3A~~ 2A Early*
~~4A~~ 3A Fannin
~~4A~~ 3A Floyd
~~4A~~ 3A Franklin
~~4A~~ 3A Gilmer
~~4A~~ 3A Gordon
~~4A~~ 3A Habersham
~~4A~~ 3A Hall
~~4A~~ 3A Lumpkin
~~4A~~ 3A Murray
~~4A~~ 3A Pickens
~~4A~~ 3A Rabun
~~4A~~ 3A Stephens
~~3A~~ 2A Tift*
~~4A~~ 3A Towns
~~4A~~ 3A Union
~~4A~~ 3A Walker
~~4A~~ 3A White
~~4A~~ 3A Whitfield
~~3A~~ 2A Worth*

HAWAII

IDAHO

ILLINOIS
~~5A~~ 4A Calhoun
~~5A~~ 4A Clark
~~5A~~ 4A Coles
~~5A~~ 4A Cumberland
~~5A~~ 4A Greene
~~5A~~ 4A Jersey

INDIANA
~~5A~~ 4A Bartholomew
~~5A~~ 4A Clay
~~5A~~ 4A Decatur
~~5A~~ 4A Fayette
~~5A~~ 4A Franklin
~~5A~~ 4A Hendricks
~~5A~~ 4A Johnson
~~5A~~ 4A Marion
~~5A~~ 4A Morgan
~~5A~~ 4A Owen
~~5A~~ 4A Putnam
~~5A~~ 4A Rush
~~5A~~ 4A Shelby
~~5A~~ 4A Union
~~5A~~ 4A Vigo

IOWA
~~6A~~ 5A Allamakee
~~6A~~ 5A Bremer
~~6A~~ 5A Buchanan
~~6A~~ 5A Buena Vista
~~6A~~ 5A Butler
~~6A~~ 5A Calhoun
~~6A~~ 5A Cherokee
~~6A~~ 5A Chickasaw
~~6A~~ 5A Clayton
~~6A~~ 5A Delaware
~~6A~~ 5A Fayette
~~6A~~ 5A Floyd
~~6A~~ 5A Franklin
~~6A~~ 5A Grundy
~~6A~~ 5A Hamilton
~~6A~~ 5A Hardin
~~6A~~ 5A Howard
~~6A~~ 5A Humboldt
~~6A~~ 5A Ida
~~6A~~ 5A Plymouth
~~6A~~ 5A Pocahontas
~~6A~~ 5A Sac
~~6A~~ 5A Webster
~~6A~~ 5A Winneshiek
~~6A~~ 5A Wright

KANSAS
~~5A~~ 4A Cloud
~~5A~~ 4A Ellis
~~5A~~ 4A Graham
~~5A~~ 4A Hamilton
~~5A~~ 4A Lane
~~5A~~ 4A Mitchell
~~5A~~ 4A Ness
~~5A~~ 4A Osborne
~~5A~~ 4A Rooks
~~5A~~ 4A Trego

(continued)

TABLE R301.1 *(continued)*

KENTUCKY

LOUISIANA

MAINE

MARYLAND
~~4A~~ 5A Allegany

MASSACHUSETTS

MICHIGAN
~~7~~ 6A Baraga
~~7~~ 6A Chippewa
~~7~~ 6A Gogebic
~~7~~ 6A Houghton
~~6A~~ 5A Huron
~~7~~ 6A Iron
~~7~~ 6A Luce
~~7~~ 6A Mackinac
~~6A~~ 7 Marquette
~~7~~ 6A Ontonagon
~~6A~~ 5A Sanilac
~~7~~ 6A Schoolcraft

MINNESOTA
~~7~~ 6A Becker
~~7~~ 6A Clay
~~6A~~ 5A Fillmore
~~7~~ 6A Grant
~~6A~~ 5A Houston
~~7~~ 6A Kanabec
~~7~~ 6A Mille Lacs
~~7~~ 6A Otter Tail
~~7~~ 6A Wilkin
~~6A~~ 5A Winona

MISSISSIPPI
~~3A~~ 2A George*

MISSOURI
~~5A~~ 4A Buchanan
~~5A~~ 4A Caldwell
~~5A~~ 4A Chariton
~~5A~~ 4A Clinton
~~4A~~ 3A Dunklin
~~4A~~ 3A Pemiscot

MONTANA

NEBRASKA

NEVADA
~~5B~~ 4B Carson City (city)
~~5B~~ 4B Douglas
~~5B~~ 4B Esmeralda
~~5B~~ 4B Lincoln
~~5B~~ 4B Lyon
~~5B~~ 4B Mineral
~~5B~~ 4B Nye

NEW HAMPSHIRE
~~6A~~ 5A Merrimack

NEW JERSEY
~~5A~~ 4A Mercer

NEW MEXICO
~~5B~~ 4B Catron
~~4B~~ 3B Sierra

NEW YORK
~~6A~~ 5A Allegany
~~6A~~ 5A Broome
~~6A~~ 5A Cattaraugus
~~6A~~ 5A Schoharie
~~6A~~ 5A Schuyler
~~6A~~ 5A Steuben
~~6A~~ 5A Tompkins
~~6A~~ 5A Wyoming

NORTH CAROLINA
~~4A~~ 3A Alamance
~~4A~~ 3A Alexander
~~4A~~ 3A Bertie
~~4A~~ 3A Caswell
~~4A~~ 3A Catawba
~~4A~~ 3A Chatham
~~4A~~ 3A Cherokee
~~4A~~ 3A Clay
~~4A~~ 3A Cleveland
~~4A~~ 3A Davie
~~4A~~ 3A Durham
~~4A~~ 3A Forsyth
~~4A~~ 3A Franklin
~~4A~~ 3A Gates
~~4A~~ 3A Granville
~~4A~~ 3A Guilford
~~4A~~ 3A Halifax
~~4A~~ 3A Harnett
~~4A~~ 3A Hertford
~~4A~~ 3A Iredell
~~4A~~ 3A Lee
~~4A~~ 3A Lincoln
~~5A~~ 4A Mitchell
~~4A~~ 3A Nash
~~4A~~ 3A Northampton
~~4A~~ 3A Orange
~~4A~~ 3A Person
~~4A~~ 3A Polk
~~4A~~ 3A Rockingham
~~4A~~ 3A Rutherford
~~4A~~ 3A Vance
~~4A~~ 3A Wake
~~4A~~ 3A Warren
~~4A~~ 3A Wilkes

NORTH DAKOTA
~~7~~ 6A Barnes
~~7~~ 6A Cass
~~7~~ 6A Eddy
~~7~~ 6A Foster
~~7~~ 6A Griggs
~~7~~ 6A Kidder
~~7~~ 6A McLean
~~7~~ 6A Mountrail
~~7~~ 6A Sheridan
~~7~~ 6A Steele
~~7~~ 6A Stutsman
~~7~~ 6A Traill
~~7~~ 6A Wells
~~7~~ 6A Williams

OHIO
~~5A~~ 4A Athens
~~5A~~ 4A Butler
~~5A~~ 4A Clinton
~~5A~~ 4A Fayette
~~5A~~ 4A Franklin
~~5A~~ 4A Greene
~~5A~~ 4A Highland
~~5A~~ 4A Hocking
~~5A~~ 4A Jackson
~~5A~~ 4A Madison
~~5A~~ 4A Meigs
~~5A~~ 4A Pickaway
~~5A~~ 4A Ross
~~5A~~ 4A Vinton
~~5A~~ 4A Warren

OKLAHOMA
~~3A~~ 4A Alfalfa
~~3A~~ 4A Craig
~~3A~~ 4A Delaware
~~3A~~ 4A Ellis
~~3A~~ 4A Garfield
~~3A~~ 4A Grant
~~3A~~ 4A Harper
~~3A~~ 4A Kay

(continued)

TABLE R301.1 (*continued*)

~~3A~~ 4A Major
~~3A~~ 4A Nowata
~~3A~~ 4A Osage
~~3A~~ 4A Ottawa
~~3A~~ 4A Washington
~~3A~~ 4A Woods
~~3A~~ 4A Woodward

OREGON

PENNSYLVANIA
~~5A~~ 4A Adams
~~5A~~ 4A Berks
~~6A~~ 5A Cameron
~~6A~~ 5A Clearfield
~~5A~~ 4A Cumberland
~~5A~~ 4A Dauphin
~~6A~~ 5A Elk
~~5A~~ 4A Franklin
~~5A~~ 4A Lancaster
~~5A~~ 4A Lebanon
~~6A~~ 5A McKean
~~5A~~ 4A Perry
~~6A~~ 5A Potter
~~6A~~ 5A Susquehanna
~~6A~~ 5A Tioga
~~6A~~ 5A Wayne

RHODE ISLAND

SOUTH CAROLINA
~~3A~~ 2A Beaufort*
~~3A~~ 2A Jasper*

SOUTH DAKOTA
~~6A~~ 5A Haakon
~~6A~~ 5A Brule
~~6A~~ 5A Jones
~~6A~~ 5A Lyman
~~6A~~ 5A Stanley

TENNESSEE
~~4A~~ 3A Bedford
~~4A~~ 3A Coffee
~~4A~~ 3A Davidson
~~4A~~ 3A Decatur
~~4A~~ 3A Franklin
~~4A~~ 3A Gibson
~~4A~~ 3A Giles
~~4A~~ 3A Grundy
~~4A~~ 3A Hamilton
~~4A~~ 3A Hickman
~~3A~~ 4A Lake
~~4A~~ 3A Lawrence
~~4A~~ 3A Lewis
~~4A~~ 3A Lincoln
~~4A~~ 3A Marion
~~4A~~ 3A Marshall
~~4A~~ 3A Maury
~~4A~~ 3A Moore
~~4A~~ 3A Perry
~~4A~~ 3A Rutherford
~~4A~~ 3A Wayne
~~4A~~ 3A Williamson

TEXAS
~~2A~~ 1A Cameron*
~~3A~~ 2A Dallas*
~~3A~~ 2A Ellis*
~~2A~~ 1A Hidalgo*
~~3A~~ 2A Johnson*
~~3A~~ 2A Navarro*
~~3A~~ 2A Tarrant*
~~2A~~ 1A Willacy*

UTAH
~~6B~~ 5B Box Elder
~~6B~~ 5B Cache
~~6B~~ 5B Carbon

VERMONT

VIRGINIA
4A (all except as follows:)
5A Alleghany
5A Bath
3A Brunswick
3A Chesapeake
5A Clifton Forge
5A Covington
3A Emporia
3A Franklin
3A Greensville
3A Halifax
3A Hampton
5A Highland
3A Isle of Wight
3A Mecklenburg
3A Newport News
3A Norfolk
3A Pittsylvania
3A Portsmouth
3A South Boston
3A Southampton
3A Suffolk
3A Surry
3A Sussex
3A Virginia Beach

WASHINGTON
~~4C~~ 5C Clallam
~~4C~~ 5C Island
~~4C~~ 5C Kitsap
~~6B~~ 5B Okanogan
~~4C~~ 5C San Juan

WEST VIRGINIA
~~5A~~ 4A Doddridge
~~5A~~ 4A Fayette
~~5A~~ 4A Greenbrier
~~5A~~ 4A Lewis
~~5A~~ 4A Nicholas
~~5A~~ 4A Raleigh
~~5A~~ 4A Summers
~~5A~~ 4A Upshur
~~5A~~ 4A Webster

WISCONSIN
~~6A~~ 5A Adams
~~7~~ 6A Ashland
~~7~~ 6A Bayfield
~~7~~ 6A Burnett
~~6A~~ 5A Calumet
~~6A~~ 5A Columbia
~~6A~~ 5A Crawford
~~6A~~ 5A Dane
~~6A~~ 5A Dodge
~~7~~ 6A Douglas
~~7~~ 6A Florence
~~6A~~ 5A Fond du Lac
~~7~~ 6A Forest
~~6A~~ 5A Grant
~~6A~~ 5A Green
~~6A~~ 5A Green Lake
~~6A~~ 5A Iowa
~~7~~ 6A Iron
~~6A~~ 5A Jefferson
~~6A~~ 5A Juneau
~~6A~~ 5A Kenosha
~~6A~~ 5A La Crosse
~~6A~~ 5A Lafayette
~~7~~ 6A Langlade
~~7~~ 6A Lincoln
~~6A~~ 5A Milwaukee
~~6A~~ 5A Monroe
~~7~~ 6A Oneida

(*continued*)

TABLE R301.1 *(continued)*

~~6A~~ 5A Outagamie
~~6A~~ 5A Ozaukee
~~7~~ 6A Price
~~6A~~ 5A Racine
~~6A~~ 5A Richland
~~6A~~ 5A Rock
~~6A~~ 5A Sauk
~~7~~ 6A Sawyer
~~7~~ 6A Taylor
~~6A~~ 5A Vernon
~~7~~ 6A Vilas
~~6A~~ 5A Walworth
~~7~~ 6A Washburn
~~6A~~ 5A Washington
~~6A~~ 5A Waukesha
~~6A~~ 5A Waushara
~~6A~~ 5A Winnebago

WYOMING
~~6B~~ 5B Laramie

US TERRITORIES
AMERICAN SAMOA

GUAM

NORTHERN MARIANA ISLANDS

PUERTO RICO
1A (all except as follows:)*
2B Barraquitas
2B Cayey

VIRGIN ISLANDS

a. Key: A – Moist, B – Dry, C – Marine. Absence of moisture designation indicates moisture regime is irrelevant.
Asterisk (*) indicates a Warm Humid location.
Note: Only counties assigned a new climate zone are shown here. Please refer to Table R301.1 for the full list of counties.

~~TABLE R301.3(1)~~ ~~International Climate Zone Definitions~~

~~MAJOR CLIMATE TYPE DEFINITIONS~~
~~Marine (C) Definition—Locations meeting all four criteria:~~ ~~1. Mean temperature of coldest month between −3°C (27°F) and 18°C (65°F).~~ ~~2. Warmest month mean < 22°C (72°F).~~ ~~3. At least four months with mean temperatures over 10°C (50°F).~~ ~~4. Dry season in summer. The month with the heaviest precipitation in the cold season has at least three times as much precipitation as the month with the least precipitation in the rest of the year. The cold season is October through March in the Northern Hemisphere and April through September in the Southern Hemisphere.~~
~~Dry (B) Definition—Locations meeting the following criteria:~~ ~~Not marine and P_{in} < 0.44 × (*TF* − 19.5) [P_{cm} < 2.0 × (*TC* + 7) in SI units]~~ ~~where:~~ ~~P_{in} = Annual precipitation in inches (cm)~~ ~~*T* = Annual mean temperature in °F (°C)~~
~~Moist (A) Definition—Locations that are not marine and not dry.~~
~~Warm-humid Definition—Moist (A) locations where either of the following wet-bulb temperature conditions shall occur during the warmest~~ **~~six consecutive months of the year:~~** ~~1. 67°F (19.4°C) or higher for 3,000 or more hours; or~~ ~~2. 73°F (22.8°C) or higher for 1,500 or more hours.~~

~~For SI: °C = [(°F) − 32]/1.8, 1 inch = 2.54 cm.~~

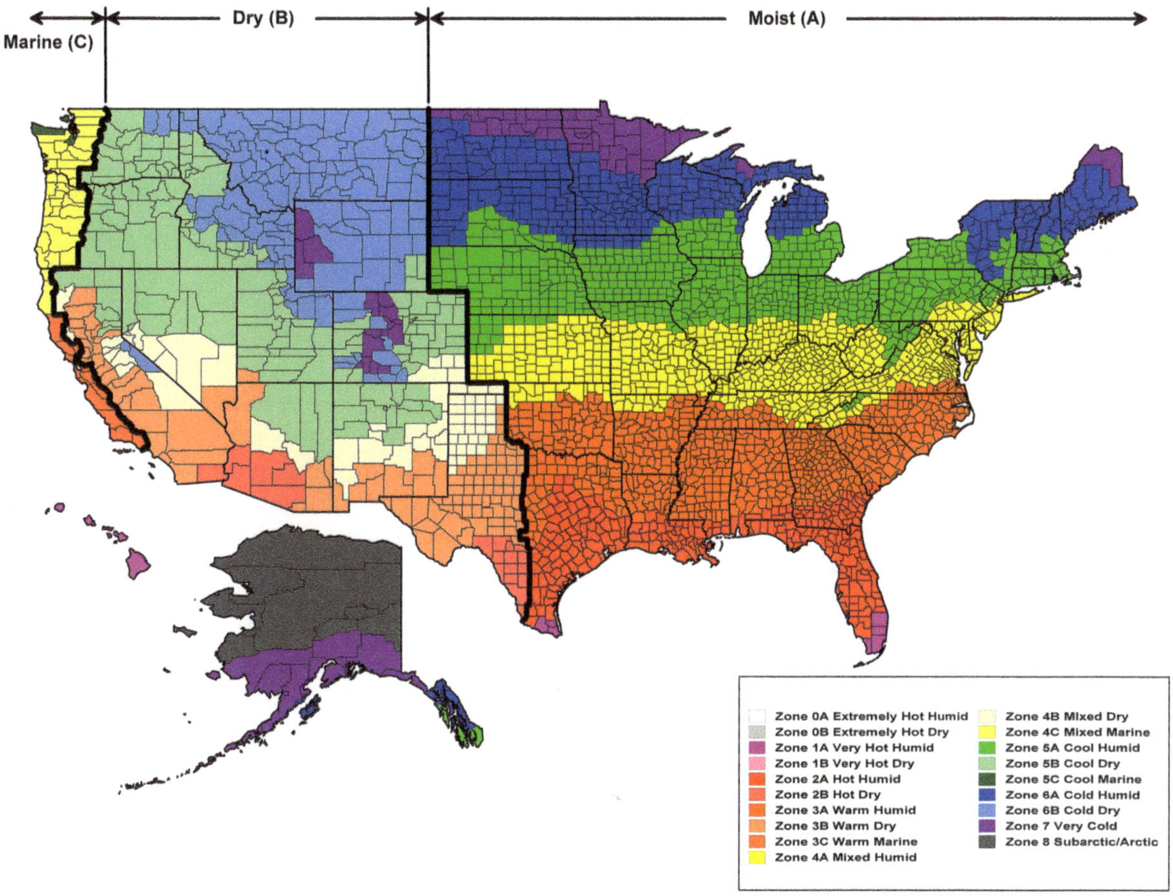

Figure R301.1 Climate Zones.

TABLE ~~R301.3(2)~~ R301.3 Thermal Climate Zone Definitions

ZONE NUMBER	THERMAL CRITERIA	
	IP Units	SI Units
0	10,800 < CDD50°F	6000 < CDD10°C
1	9,000 < CDD50°F < 10,800	5000 < CDD10°C < 6000
2	6,300 < CDD50°F ≤ 9,000	3500 < CDD10°C ≤ 5000
~~3A and 3B~~ 3	~~4,500 <~~ CDD50°F ≤ 6,300 AND HDD65°F ≤ ~~5,400~~ 3,600	CDD10°C < 3500 AND HDD18°C ≤ ~~3000~~ 2000
~~4A and 4B~~ 4	CDD50°F ≤ ~~4,500~~ 6,300 AND 3,600 < HDD65°F ≤ 5,400	CDD10°C < ~~2500~~ 3500 AND 2000 < HDD18°C ≤ 3000
~~3C~~	~~HDD65°F ≤ 3,600~~	~~HDD18°C ≤ 2000~~
~~4C~~	~~3,600 < HDD65°F ≤ 5,400~~	~~2000 < HDD18°C ≤ 3000~~
5	CDD50°F < 6,300 AND 5,400 < HDD65°F ≤ 7,200	CDD10°C < 3500 AND 3000 < HDD18°C ≤4000
6	7,200 < HDD65°F ≤ 9,000	4000 < HDD18°C ≤5000
7	9,000 < HDD65°F ≤ 12,600	5000 < HDD18°C ≤ 7000
8	12,600 < HDD65°F	7000 < HDD18°C

For SI: °C = [(°F) − 32]/1.8.

CHANGE SIGNIFICANCE: The climate zone provisions in Chapter 3, from Section R301.3 through the climate zone tables and map, have undergone a significant update. This update corresponds with ASHRAE Standard 169-2013, which is referenced in the 2018 IgCC, ASHRAE 90.1 and ASHRAE 90.2. ICC has a licensing agreement with ASHRAE to include the climate zone map, definitions and tables for consistency with ASHRAE Standard 169 and other key codes and standards.

These two key updates are relevant to most code users:

- Updates to the climate zone designations in Table R301.1 for approximately 10 percent of the counties in the United States. In general, the counties were assigned a warmer climate zone.
- Updates to Figure R301.1, the climate zone map, to reflect the changes in county climate zone designations.

The number of climate zones increased to include Climate Zone 0, representing the hottest of climates. Climate Zone 0 is a subset of the previous Climate Zone 1, which included all locations with more than 9,000 Cooling Degree Days (CDDs). Climate Zone 1 now tops out at 10,800 CDDs, and Climate Zone 0 is for those locations with more. Cities in Climate Zone 0 include extremely hot locations such as Mumbai (Bombay), Jakarta and Abu Dhabi. There are no cities in the United States in Climate Zone 0. Several code changes introduce specific provisions for Climate Zone 0. Because Climate Zone 0 is a subset of the former Climate Zone 1, references throughout the IECC to CZ 1 have been edited to CZ 0 and 1.

Code users needing to calculate the moisture zone (marine, dry or humid) will find Section R301.3 entirely revised. Calculations are in three categories: (a) when monthly average temperature and precipitation data are available, (b) when annual average temperature, degree days and precipitation data are available, and (c) when only degree day information is available.

R301.3(1) has been eliminated and R301.3(2) renumbered to R301.3. Section R301.3 now includes Climate Zone 0 and the revised heating and cooling degree day thresholds for climate zones. Table R301.3 Thermal Climate Zone Definitions has also been revised to include Climate Zone 0 and the adjusted degree days for other climate zones.

PART 7

Residential Energy Provisions: Energy Efficiency Requirements

Chapters 4 and 5

- **Chapter 4** Residential Energy Efficiency
- **Chapter 5** Existing Buildings
  No changes addressed

Chapters 4 and 5 of the *International Energy Conservation Code* (IECC) contain the bulk of the design and construction provisions for residential buildings. The definition of a *residential building* is unique to this code. The IECC defines *residential buildings* as detached one- and two-family dwellings and multiple single-family dwellings as well as R-2, R-3 or R-4 buildings three stories or less in height. All other buildings are regulated by the energy conservation requirements in the IECC—Commercial provisions. The applicable portions of a residential building must comply with the provisions in chapters 4 and 5 for energy efficiency.

Chapter 4 defines the requirements for the entire building system, including the building envelope and mechanical, service water heating and electrical power and lighting systems. Chapter 4 also includes required additional efficiency options and alternative compliance paths.

Chapter 5 addresses construction on existing buildings – additions, alterations and changes to occupancy – correlating to the requirements of Chapter 4. ■

R401.2
Compliance Paths

R401.3
Certificate

R402
Building Envelope

R402.1.2
Insulation and Fenestration Criteria

TABLE R402.1.2
Insulation and Fenestration *U*-Factors

TABLE R402.1.3
Insulation Minimum *R*-Values and Fenestration Requirements by Component

R402.2.8
Basement Walls

R402.4.1.2
Air Leakage Testing

R402.5
Maximum Fenestration *U*-Factor and SHGC

R403
Systems

R403.3.1-R403.3.3.1

Duct Location and Insulation

R403.3.5

Duct Testing

R403.6.3

Mechanical Ventilation System Testing

R404

Electrical Power and Lighting Systems

R404.1.1

Exterior Lighting

R404.2

Interior Lighting Controls

R404.3

Exterior Lighting Controls

R405

Total Building Performance

R405.2

Performance-Based Compliance

R405.3.2

Compliance Report

R406

Energy Rating Index Compliance Alternative

R406.2

ERI Compliance

R406.3

Building Thermal Envelope

406.7.3

Renewable Energy Certificate Documentation

407

Tropical Climate Region

R408

Additional Efficiency Package Options

R401.2

Compliance Paths

CHANGE TYPE: Modification

CHANGE SUMMARY: Compliance path options for the residential provisions of the IECC are named and the sections required for each option are outlined.

2021 CODE: **R401.2 ~~Compliance~~ Application.** ~~Projects~~ Residential buildings shall comply with Section R401.2.5 and either Sections R401.2.1, R401.2.2, R401.2.3 or R401.2.4. ~~one of the following~~:

~~1. Sections R401 through R404.~~

~~2. Section R405 and the provisions of Sections R401 through R404 indicated as "Mandatory."~~

~~3. The energy rating index (ERI) approach in Section R406.~~

Exception: Additions, alterations, repairs and changes of occupancy to existing buildings complying with Chapter 5.

R401.2.1 Prescriptive Compliance Option. The Prescriptive Compliance Option requires compliance with Sections R401 through R404.

Residential projects must comply with one of four compliance options.

R401.2.2 Total Building Performance Option. The Total Building Performance Option requires compliance with Section R405~~, and the provisions of Sections R401 through R404 indicated as "Mandatory".~~

R401.2.3 Energy Rating Index Option. The Energy Rating Index (ERI) Option requires compliance with Section R406.

R401.2.4 Tropical Climate Region Option. The Tropical Climate Region option requires compliance with Section R407.

R401.2.5 Additional energy efficiency. This section establishes additional requirements applicable to all compliance approaches to achieve additional energy efficiency.

1. For buildings complying with Section R401.2.1, one of the additional efficiency package options shall be installed according to Section R408.2.
2. For buildings complying with Section R401.2.2, the building shall meet one of the following:
   - **2.1.** One of the additional efficiency package options in Section R408.2 shall be installed without including such measures in the proposed design under Section R405; or
   - **2.2.** The proposed design of the building under Section R405.3 shall have an annual energy cost that is less than or equal to 95 percent of the annual energy cost of the standard reference design.
3. For buildings complying with the Energy Rating Index alternative Section R401.2.3, the Energy Rating Index value shall be at least 5 percent less than the Energy Rating Index target specified in Table R406.5.

The option selected for compliance shall be identified in the certificate required by Section R401.3.

CHANGE SIGNIFICANCE: Beginning in the 2006 IECC, Section R401.2 identified two options for compliance with the technical provisions included in Chapter 4 of the IECC. The options are not formally named within the code but have commonly been referred to as the prescriptive and performance compliance paths. A third compliance path, the Energy Rating Index, was added to the 2015 IECC along with a simplified compliance approach for certain dwelling units located in the equatorial Tropical Region; the latter was not listed with the other compliance paths in Section R401.2.

Section R401.2 has been revised to clarify and name the four compliance paths for residential projects. Naming the compliance options (Prescriptive, Total Building Performance, Energy Rating Index and Tropical Climate Region) formalizes the way in which the paths are identified and adds clarity for training, education and new code users. Referring to the four paths as options reinforces that it is the designer's choice as to which path is followed.

The terms *prescriptive* and *mandatory* in provision titles throughout Chapter 4 are also eliminated, reducing confusion about the application of the code requirements labeled as such. This corresponds to the creation of project requirement tables in Sections R405 and R406; all requirements associated with these compliance paths are now contained within the respective sections.

Additionally, the requirements for Tropical Climate Region compliance have been relocated to the new Section R407.

R401.3

Certificate

CHANGE TYPE: Modification

CHANGE SUMMARY: Requirements for the certificate are expanded to include the code edition and compliance path for the project as well as information related to photovoltaic systems and Energy Rating Index scores, if applicable.

Energy Efficiency Certificate

Code edition	
Compliance path	

Insulation Rating		*R*-Value		*R*-Value
Ceiling/Roof		*R*-		*R*-
Walls	Frame	*R*-	Mass	*R*-
	Basement	*R*-	Crawl space	*R*-
Floors	Over unconditioned space	*R*-	Slab edge	*R*-
Ducts	Attic	*R*-	Other	*R*-

Air Leakage Test Results

Envelope testing		**ACH**	______**Pa.**	**Duct testing**		**cfm/100 ft²**

Fenestration Rating	NFRC *U*-Factor	NFRC SHGC
Window	*U*-	
Opaque door	*U*-	
Skylight	*U*-	
Weighted average	*U*-	

Equipment Performance	Type	Efficiency		Size
Heating system			AFUE	
Cooling system			SEER	
Water heater			EF	

Indicate if the following have been installed (an efficiency shall not be listed)

☐ electric furnace ☐ gas-fire unvented room heater ☐ baseboard electric heater

Additional Energy Efficiency (check one)

☐ **Proposed design had an annual energy cost ≤ 95% of that of the reference design**

☐ **Energy Rating Index score is at least 5% less than ERI target**

☐ **Additional efficiency package option is installed (specify option)**

Photovoltaic Panel System		Energy Rating Index Score	
Array capacity		**with PV**	
Inverter efficiency		**without PV**	
Panel tilt			
Orientation			

Designer/builder ____________ **Date** ____________

This Certificate is to be posted in accordance with Section R401.3 of the International Energy Conservation Code.

The Certificate must now include compliance path, among other new items.

2021 CODE: **R401.3 Certificate ~~(Mandatory)~~.** A permanent certificate shall be completed by the builder or other approved party and posted on a wall in the space where the furnace is located, a utility room or an approved location inside the building. Where located on an electrical panel, the certificate shall not cover or obstruct the visibility of the circuit directory label, service disconnect label or other required labels. The certificate shall indicate the following:

1. The predominant *R*-values of insulation installed in or on ceilings, roofs, walls, foundation components such as slabs, basement walls, crawl space walls and floors and ducts outside conditioned spaces.
2. *U*-factors of fenestration and the solar heat gain coefficient (SHGC) of fenestration. Where there is more than one value for ~~each~~ any component of the building envelope, the certificate shall indicate both the value covering the largest area and the area weighted average value if available.
3. The results from any required duct system and building envelope air leakage testing performed on the building.
4. The types, sizes and efficiencies of heating, cooling and service water-heating equipment. Where a gas-fired unvented room heater, electric furnace or baseboard electric heater is installed in the residence, the certificate shall indicate "gas-fired unvented room heater," "electric furnace" or "baseboard electric heater," as appropriate. An efficiency shall not be indicated for gas-fired unvented room heaters, electric furnaces and electric baseboard heaters.
5. Where onsite photovoltaic panel systems have been installed, the array capacity, inverter efficiency, panel tilt and orientation shall be noted on the certificate.
6. For buildings where an Energy Rating Index score is determined in accordance with Section R406, the Energy Rating Index score, both with and without any on-site generation, shall be listed on the certificate.
7. The code edition under which the structure was permitted, and the compliance path used.

CHANGE SIGNIFICANCE: The updates to the certificate reflect changes made to the IECC in recent code cycles and include information that is beneficial for compliance purposes and for future homeowners performing maintenance or replacing equipment. In addition to detail on thermal envelope efficiency and HVAC equipment size, the certificate now requires the compliance path and code version under which the project was permitted, details on renewable energy systems installed and information related to the Energy Rating Index score, if applicable. The information required for the certificate is readily available at construction and on project documentation, and it will take very little effort to transfer it onto the permanent certificate. However, such information may be difficult or impossible to determine in the future.

This provision is applicable to all compliance paths.

R402

Building Envelope

Section R402 provides building thermal envelope provisions for residential buildings, including limitations on air infiltration, or air leakage. The building envelope is important to building energy efficiency. When it is cold outside, heat loss and air leakage through the building envelope add to the heating load. On hot days, solar gains through windows, heat gain through opaque assemblies and the infiltration of hot or humid air contribute to the cooling load. The building envelope requirements of Section R402 are intended to reduce these heat gains and losses through the building envelope.

The residential envelope provisions have seen few changes since the 2012 IECC. Significant changes to the building envelope in the 2021 IECC include:

- Increased insulation requirements
- Reduced fenestration *U*-factors and solar heat gain coefficients (SHGCs)
- Increased air leakage requirements
- Revised duct testing requirements.

Photo courtesy of c12/iStock/Getty Images Plus

Efficient building envelopes reduce heat gains and losses.

R402.1.2

Insulation and Fenestration Criteria

CHANGE TYPE: Modification

CHANGE SUMMARY: The assembly *U*-Factor is established as the primary insulation metric, and *R*-Value is an alternative.

2021 CODE: **R402.1.2 Insulation and fenestration criteria.** The building thermal envelope shall meet the requirements of Table R402.1.2, based on the climate zone specified in Chapter 3. Assemblies shall have a *U*-factor equal to or less than that specified in Table R402.1.2. Fenestration shall have a *U*-factor and glazed fenestration SHGC equal to or less than that specified in Table R402.1.2.

R402.1.3 R-value alternative. Assemblies with *R*-value of insulation materials equal to or greater than that specified in Table R402.1.3 shall be an alternative to the *U*-factor in Table R402.1.2.

CHANGE SIGNIFICANCE: Since the 2006 IECC, *R*-values have been the primary metric for envelope efficiencies, with assembly *U*-factors serving as an alternative compliance metric. The 2021 IECC focuses envelope efficiency on *U*-factors which define the performance of entire assemblies. The *R*-value approach is retained, providing predetermined solutions based on the *U*-factor requirements. This reorientation ensures the *R*-values used are a derivative of the intended performance levels represented by the assembly *U*-factors. The SHGC values (and footnotes) are consistent with those in the previous *R*-value table.

These provisions are applicable to the Prescriptive Compliance path.

Photo courtesy of Fred de Noyelle/Stone/Getty Images

U-factor is used to measure thermal transmittance of fenestration and envelope assemblies.

Table R402.1.2

Insulation and Fenestration *U*-Factors

CHANGE TYPE: Modification

CHANGE SUMMARY: The ceiling and framed wall assembly *U*-factors in Table R402.1.2 have been decreased.

2021 CODE: Table R402.1.2.

CHANGE SIGNIFICANCE: The 2021 IECC increases efficiencies for ceilings, wood frame walls and fenestration. Specifically, the ceiling *U*-factors for Climate Zones 2 and 3 are decreased to align with the 2018 IECC requirements for Climate Zone 4 and higher; similarly, ceiling *U*-factors are decreased for Climate Zones 4 through 8, increasing efficiency beyond what was achieved in previous code versions. The *U*-factor requirements for wood frame walls in Climate Zones 4 and 5 are decreased as well, aligning with the 2018 IECC requirements for Climate Zones 6 through 8. Improved efficiencies in fenestration *U*-factors for Climate Zones 3 and 4 are also achieved. Finally, SHGC values are copied from the *R*-value table to the *U*-factor table, with requirements added for Climate Zone 5 that are consistent with the 2018 IECC requirements for Climate Zone 4 (0.40).

Table R402.1.2 introduces a footnote that provides an exception to the prescriptive *U*-factors for fenestration installed at high altitudes (above 4,000 feet in elevation) and in regions that require fenestration to be resistant to windborne debris in Climate Zones 3 through 8.

This table is applicable to the Prescriptive compliance path.

Photo courtesy of BanksPhotos/E+/Getty Images

U-factors have been decreased for greater efficiency.

TABLE ~~R402.1.4~~ R402.1.2 ~~Equivalent~~ Maximum Assembly *U*-Factors[a] and Fenestration Requirements

CLIMATE ZONE	FENESTRATION *U*-FACTOR[f]	SKYLIGHT *U*-FACTOR	GLAZED FENESTRATION SHGC[d,e]	CEILING *U*-FACTOR	WOOD FRAME WALL *U*-FACTOR	MASS WALL *U*-FACTOR[b]	FLOOR *U*-FACTOR	BASEMENT WALL *U*-FACTOR	CRAWL SPACE WALL *U*-FACTOR
0	0.50	0.75	0.25	0.035	0.084	0.197	0.064	0.360	0.477
1	0.50	0.75	0.25	0.035	0.084	0.197	0.064	0.360	0.477
2	0.40	0.65	0.25	~~0.030~~ 0.026	0.084	0.165	0.064	0.360	0.477
3	~~0.32~~ .30	0.55	0.25	~~0.030~~ 0.026	0.060	0.098	0.047	0.091[c]	0.136
4 except Marine	~~0.32~~ .30	0.55	0.40	~~0.026~~ 0.024	~~0.060~~ 0.045	0.098	0.047	0.059	0.065
5 and Marine 4	0.30	0.55	0.40	~~0.026~~ 0.024	~~0.060~~ 0.045	0.082	0.033	0.050	0.055
6	0.30	0.55	NR	~~0.026~~ 0.024	0.045	0.060	0.033	0.050	0.055
7 and 8	0.30	0.55	NR	~~0.026~~ 0.024	0.045	0.057	0.028	0.050	0.055

For SI: 1 foot = 304.8 mm.

a. Nonfenestration *U*-factors shall be obtained from measurement, calculation or an approved source.

b. Mass walls shall be in accordance with Section R402.2.5. Where more than half the insulation is on the interior, the mass wall *U*-factors shall not exceed 0.17 in Climate Zones 0 and 1, 0.14 in Climate Zone 2, 0.12 in Climate Zone 3, 0.087 in Climate Zone 4 except Marine, 0.065 in Climate Zone 5 and Marine 4, and 0.057 in Climate Zones 6 through 8.

c. In Warm Humid locations as defined by Figure R301.1 and Table R301.1, the basement wall *U*-factor shall not exceed 0.360.

d. The SHGC column applies to all glazed fenestration.

Exception: In Climate Zones 0 through 3, skylights shall be permitted to be excluded from glazed fenestration SHGC requirements provided that the SHGC for such skylights does not exceed 0.30.

e. There are no SHGC requirements in the Marine Zone.

f. A maximum *U*-factor of 0.32 shall apply in Marine Climate Zone 4 and Climate Zones 5 through 8 to vertical fenestration products installed in buildings located either:

1. Above 4,000 feet in elevation above sea level, or
2. In windborne debris regions where protection of openings is required by Section R301.2.1.2 of the International Residential Code.

Table R402.1.3

Insulation Minimum *R*-Values and Fenestration Requirements by Component

CHANGE TYPE: Modification

CHANGE SUMMARY: The *R*-value table has been restructured and the efficiencies for most components have been increased.

2021 CODE: Table R402.1.3.

CHANGE SIGNIFICANCE: Efficiencies are increased for ceilings, wood frame walls, basement walls, crawlspace walls, slabs on grade and fenestration. Ceiling *R*-value requirements for Climate Zones 2 and 3 are increased to align with the 2018 IECC requirements for Climate Zone 4 and higher, and the efficiencies in Climate Zones 4 through 8 are similarly increased.

Insulation *R*-values for wood frame walls increased in Climate Zones 4 and 5, aligning with the 2018 IECC requirements for Climate Zone 6.

Basement and crawlspace wall requirements are amended for clarity and the alternative compliance option (13 & 5ci) is moved from the footnote to the table. The clarification is made by adding a "ci" designator for the continuous insulation components in the table. This is also carried through in footnote c. The footnote is further clarified by consistently referring to "wall" rather than "basement" or "home" with no mention of "crawlspace." The footnote applies to basement and crawlspace walls as noted in the heading of the table columns. The word "surface" is added to clarify that where continuous insulation is used, it should be applied to the surface of the foundation wall (and any additional cavity insulation applied to the interior side of the continuous insulation).

Slab on grade insulation efficiency has been increased for Climate Zones 3 through 5. The slab *R*-value requirements had not been improved in any climate zone since the 2006 IECC or before.

Photo courtesy of geremme/E+/Getty Images

R-values have been increased for greater efficiency.

~~TABLE R402.1.2~~ TABLE R402.1.3 Insulation Minimum *R*-values and Fenestration Requirements By Component[a]

CLIMATE ZONE	FENESTRATION *U*-FACTOR[b,i]	SKYLIGHT[b] *U*-FACTOR	GLAZED FENESTRATION SHGC[b,e]	CEILING *R*-VALUE	WOOD FRAME WALL *R*-VALUE[g]	MASS WALL *R*-VALUE[h]	FLOOR *R*-VALUE	BASEMENT[c,g] WALL *R*-VALUE	SLAB[d] *R*-VALUE & DEPTH	CRAWL SPACE[c,g] WALL *R*-VALUE
0	NR	0.75	0.25	30	13 or 0 & 10ci[g]	3/4	13	0	0	0
1	NR	0.75	0.25	30	13 or 0 & 10ci[g]	3/4	13	0	0	0
2	0.40	0.65	0.25	~~38~~ 49	13 or 0 & 10ci[g]	4/6	13	0	0	0
3	~~0.32~~ .30	0.55	0.25	~~38~~ 49	20 or ~~13+5[h]~~ 13 & 5ci[h] or 0 & 15ci[g,h]	8/13	19	~~5/13[f]~~ 5ci or 13[f]	~~0~~10ci, 2 ft	~~5/13~~ 5ci or 13[f]
4 except Marine	~~0.32~~ .30	0.55	0.40	~~49~~ 60	30 or 20 & 5ci[h] or 13 & 10ci[h] or 0 & 20ci[h]	8/13	19	~~10 /13~~ 10ci or 13	10ci, ~~2~~ 4 ft	~~10/13~~ 10ci or 13
5 and Marine 4	0.30[i]	0.55	~~NR~~ 0.40	~~49~~ 60	30 or 20 & 5ci[h] or 13 & 10ci[h] or 0 & 20ci[h]	13/17	30[g]	~~15/19~~ 15ci or 19 or 13 & 5ci	10ci, ~~2~~ 4 ft	~~15/19~~ 15ci or 19 or 13 & 5ci
6	0.30[i]	0.55	NR	~~49~~ 60	30 or 20 & 5ci[h] or 13 & 10ci[h] or 0 & 20ci[h]	15/20	30[g]	~~15/19~~ 15ci or 19 or 13 & 5ci	10ci, 4 ft	~~15/19~~ 15ci or 19 or 13 & 5ci
7 and 8	0.30[i]	0.55	NR	~~49~~ 60	30 or 20 & 5ci[h] or 13 & 10ci[h] or 0 & 20ci[h]	19/21	38[g]	~~15/19~~ 15ci or 19 or 13 & 5ci	10ci, 4 ft	~~15/19~~ 15ci or 19 or 13 & 5ci

For SI: 1 foot = 304.8 mm.

NR = Not Required.

ci = continuous insulation.

a. *R*-values are minimums. *U*-factors and SHGC are maximums. Where insulation is installed in a cavity that is less than the label or design thickness of the insulation, the installed *R*-value of the insulation shall be not less than the *R*-value specified in the table.

b. The fenestration *U*-factor column excludes skylights. The SHGC column applies to all glazed fenestration.

Exception: In Climate Zones 0 through 3, skylights shall be permitted to be excluded from glazed fenestration SHGC requirements provided that the SHGC for such skylights does not exceed 0.30.

c. "5ci or 13" means R-5 continuous insulation (ci) on the interior or exterior surface of the wall or R-13 cavity insulation on the interior of the wall.
"10ci or 13" means R-10 continuous insulation (ci) on the interior or exterior surface of the wall or R-13 cavity insulation on the interior of the wall.
"15ci or 19 or 13 & 5ci" means R-15 continuous insulation (ci) on the interior or exterior surface of the wall; or R-19 cavity insulation on the interior of the wall; or R-13 cavity insulation on the interior of the wall in addition to R-5 continuous insulation on the interior or exterior surface of the wall.

d. R-5 insulation shall be provided under the full slab area of a heated slab in addition to the required slab edge insulation *R*-value for slabs, as indicated in the table. The slab-edge insulation for heated slabs shall not be required to extend below the slab.

e. There are no SHGC requirements in the Marine Zone.

f. Basement wall insulation is not required in Warm Humid locations as defined by Figure R301.1 and Table R301.1.

~~g. Alternatively, insulation sufficient to fill the framing cavity and providing not less than an *R*-value of R-19.~~

g. The first value is cavity insulation; the second value is continuous insulation. Therefore, as an example, "13 & 5" means R-13 cavity insulation plus R-5 continuous insulation.

h. Mass walls shall be in accordance with Section R402.2.5. The second *R*-value applies where more than half of the insulation is on the interior of the mass wall.

i. A maximum *U*-factor of 0.32 shall apply in Climate Zones 3 through 8 to vertical fenestration products installed in buildings located either:

1. Above 4,000 feet in elevation above sea level, or
2. In windborne debris regions where protection of openings is required by Section R301.2.1.2 of the International Residential Code.

Floor insulation footnote g in Climate Zones 4 through 8 improves the efficiency of dwellings in the coldest climate zones by removing an exception that allows weaker floor insulation *R*-values with no corresponding improvements elsewhere in the building.

Fenestration *U*-factors in Climate Zones 3 and 4 are similarly increased in efficiency, and SHGC has been introduced in Climate Zone 5 where it was previously not regulated.

Finally, Table R402.1.2 introduces a footnote that provides an exception to the prescriptive *U*-factors for fenestration installed at high altitudes (above 4,000 feet in elevation) and in regions that require fenestration to be resistant to windborne debris in Climate Zones Marine 4 and 5 through 8.

This table is applicable to the Prescriptive compliance path.

R402.2.8

Basement Walls

CHANGE TYPE: Modification

CHANGE SUMMARY: Exceptions to the basement walls provision clearly define the characteristics of unconditioned basements.

2021 CODE: **R402.2.8 Basement walls.** Basement walls shall be insulated in accordance with Table R402.1.3.

Exception: Basement walls associated with unconditioned basements where all of the following requirements are met:

1. The floor overhead, including the underside stairway stringer leading to the basement, is insulated in accordance with Section R402.1.3 and applicable provisions of Sections R402.2 and R402.2.7.
2. There are no uninsulated duct, domestic hot water, or hydronic heating surfaces exposed to the basement.
3. There are no HVAC supply or return diffusers serving the basement.
4. The walls surrounding the stairway and adjacent to conditioned space are insulated in accordance with Section R402.1.3 and applicable provisions of Section R402.2.
5. The door(s) leading to the basement from conditioned spaces are insulated in accordance with Section R402.1.3 and applicable provisions of Section R402.2, and weatherstripped in accordance with Section R402.4.
6. The building thermal envelope separating the basement from adjacent conditioned spaces complies with Section R402.4.

Photo courtesy of slobo/E+/Getty Images

Unconditioned basements are exempt from basement wall provisions.

R402.2.8.1 ~~R402.2.9~~ Basement wall insulation installation. ~~Walls associated with conditioned basements~~ Where basement walls are insulated, the insulation shall be ~~insulated~~ installed from the top of the basement wall down to 10 feet (3048 mm) below grade or to the basement floor, whichever is less.

CHANGE SIGNIFICANCE: Basement walls are required to comply with the provisions in Table R402.1.3 unless they meet each of the characteristics for unconditioned basements. Previous versions of the energy code did not consider the heat transfer through ducts, stairways, doors and other elements of the basement when defining an unconditioned basement. Changes to this section clearly outline the characteristics that must be present in an unconditioned basement to remove the basement from consideration as an extension of conditioned space.

These provisions are applicable to the Prescriptive compliance path.

R402.4.1.2

Air Leakage Testing

CHANGE TYPE: Modification

CHANGE SUMMARY: The air leakage requirements are revised and a new definition for dwelling unit enclosure area is provided.

2021 CODE: **DWELLING UNIT ENCLOSURE AREA.** The sum of the area of ceiling, floors and walls separating a dwelling unit's conditioned space from the exterior or from adjacent conditioned or unconditioned spaces. Wall height shall be measured from the finished floor of the dwelling unit to the underside of the floor above.

R402.4.1.2 Testing. The building or dwelling unit shall be tested ~~and verified as having an air leakage rate not exceeding five 5.0 air changes per hour in Climate Zones 1 and 2, and three 3.0 air changes per hour in Climate Zones 3 through 8.~~ for air leakage. The maximum air leakage rate for any building or dwelling unit under any compliance path shall not exceed 5.0 air changes per hour or 0.28 cubic feet per minute (CFM) per square foot [$0.0079\ m^3/(s \times m^2)$] of dwelling unit enclosure area. Testing shall be conducted in accordance with ANSI/RESNET/ICC 380, ASTM E779 or ASTM E1827 and reported at a pressure of 0.2 inch w.g. (50 Pascals). Where

Calibrated blower door test

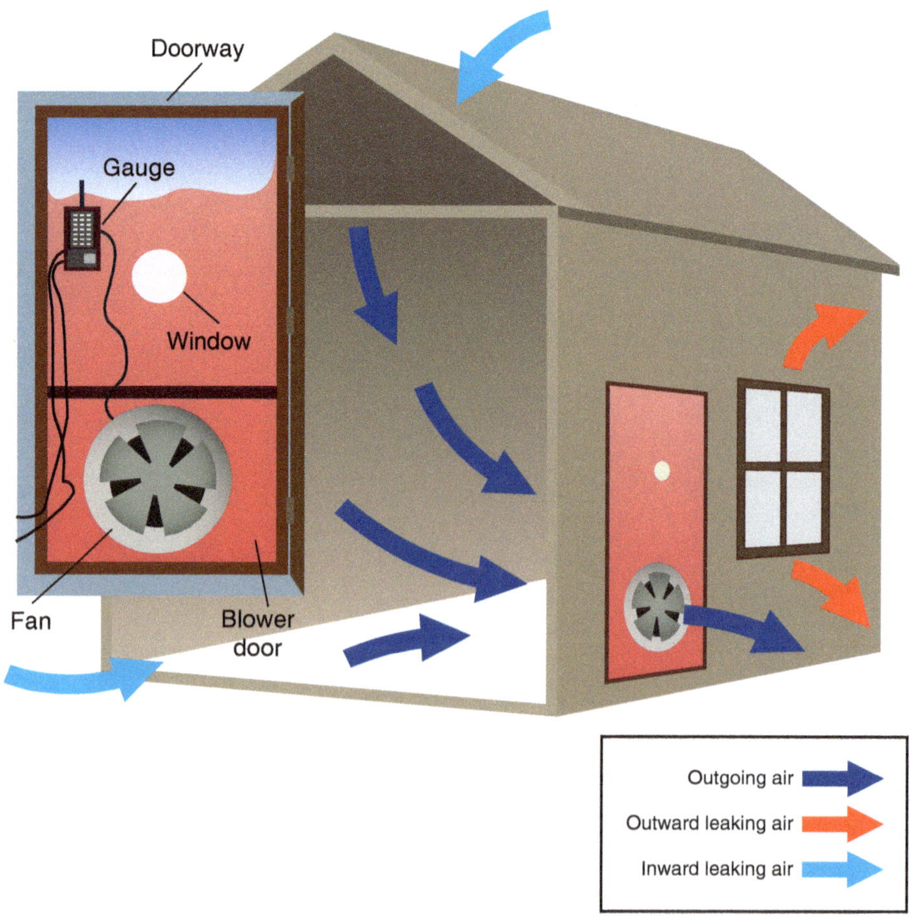

Photo courtesy of Building Performance Institute Inc.

Air leakage testing is required for residential projects.

required by the code official, testing shall be conducted by an approved third party. A written report of the results of the test shall be signed by the party conducting the test and provided to the code official. Testing shall be performed at any time after creation of all penetrations of the building thermal envelope have been sealed.

> **Exception:** For heated, attached private garages and heated, detached private garages accessory to one- and two-family dwellings and townhouses not more than three stories above grade plane in height, building envelope tightness and insulation installation shall be considered acceptable where the items in Table R402.4.1.1, applicable to the method of construction, are field verified. Where required by the code official, an approved third party independent from the installer shall inspect both air barrier and insulation installation criteria. Heated, attached private garage space and heated, detached private garage space shall be thermally isolated from all other habitable, conditioned spaces in accordance with Sections R402.2.12 and R402.3.5, as applicable.

During testing:

1. Exterior windows and doors, fireplace and stove doors shall be closed, but not sealed, beyond the intended weatherstripping or other infiltration control measures.
2. Dampers including exhaust, intake, makeup air, backdraft and flue dampers shall be closed, but not sealed beyond intended infiltration control measures.
3. Interior doors, where installed at the time of the test, shall be open.
4. Exterior or interior terminations for continuous ventilation systems shall be sealed.
5. Heating and cooling systems, where installed at the time of the test, shall be turned off.
6. Supply and return registers, where installed at the time of the test, shall be fully open.

> **Exception:** When testing individual dwelling units, an air leakage rate not exceeding 0.30 cubic feet per minute per square foot [0.008 $m^3/(s \times m^2)$] of the dwelling unit enclosure area, tested in accordance with ANSI/RESNET/ICC 380, ASTM E779 or ASTM E1827 and reported at a pressure of 0.2 inch w.g. (50 Pa), shall be permitted in all climate zones for:
>
> 1. Attached single and multiple-family building dwelling units.
> 2. Buildings or dwelling units that are 1,500 square feet (139.4 m^2) or smaller.

Mechanical ventilation shall be provided in accordance with Section M1505 of the International Residential Code or Section 403.3.2 of the International Mechanical Code, as applicable, or with other approved means of ventilation.

R402.4.1.3 Leakage rate. When complying with Section R401.2.1, the building or dwelling unit shall have an air leakage rate not exceeding 5.0 air changes per hour in Climate Zones 0, 1 and 2, and 3.0 air changes per hour in Climate Zones 3 through 8, when tested in accordance with Section R402.4.1.2.

CHANGE SIGNIFICANCE: Several significant changes have been made to the testing requirements for residential projects. The maximum air leakage rate is now 5 air changes per hour (ACH) for all compliance paths and 3 ACH for dwellings in Climate Zones 3 through 8 following the Prescriptive compliance path. The air leakage rate for buildings and dwelling units following the Prescriptive compliance path is detailed in the new Section R402.4.1.3. Heated attached and detached garages that are field verified for air barrier and insulation installation criteria in Table R402.4.1.1 are exempt from the testing requirements.

An alternative testing metric has been added to the 2021 IECC for attached single and multiple-family dwelling units and smaller homes. While median-sized single-family homes are typically able to achieve the air leakage rate required by the IECC, it can be more difficult for small volume or attached homes to comply. The alternative metric more accurately reflects leakage through the exterior enclosure area which ensures even small volume homes are able to comply with the required air leakage rates.

These provisions are applicable to all compliance paths except the Tropical Climate Region compliance option.

R402.5

Maximum Fenestration *U*-Factor and SHGC

CHANGE TYPE: Modification

CHANGE SUMMARY: The maximum fenestration *U*-factors are reduced, and an exception is provided for storm shelters.

2021 CODE: R402.5 Maximum fenestration *U*-factor and SHGC (~~Mandatory~~). The area-weighted average maximum fenestration *U*-factor permitted using tradeoffs from Section R402.1.5 or R405 shall be 0.48 in Climate Zones 4 and 5 and 0.40 in Climate Zones 6 through 8 for vertical fenestration, and 0.75 in Climate Zones 4 through 8 for skylights. The area-weighted average maximum fenestration SHGC permitted using tradeoffs from Section R405 in Climate Zones 0 through 3 shall be ~~0.50~~ 0.40.

> **Exception:** The maximum *U*-factor and solar heat gain coefficient (SHGC) for fenestration shall not be required in storm shelters complying with ICC 500.

CHANGE SIGNIFICANCE: Maximum fenestration *U*-factors have been in the IECC since the 2006 version and serve as a backstop in the event of tradeoffs. Over four code update cycles, from the 2009 IECC through the 2018 IECC, prescriptive fenestration efficiencies increased, but the maximum *U*-factors and SHGCs used for tradeoffs were not updated. To remain consistent with improvements made to the prescriptive fenestration values, maximum fenestration *U*-factors and SHGCs permitting in the tradeoff compliance options are updated in the 2021 IECC.

This provision is applicable to the Prescriptive compliance option where tradeoffs are used and to the Total Building Performance compliance option.

Photo courtesy of KatarzynaBialasiewicz/iStock/Getty Images Plus

U-Factor and SHGC requirements indicate heat transmittance of fenestration.

R403

Systems

Section R403 broadly addresses the efficiency, construction and operation of building mechanical systems that are not regulated by the U.S. Department of Energy. Federal regulation establishes energy efficiency standards for many consumer and commercial appliances under the Energy Policy and Conservation Act (EPCA). States are preempted from creating energy efficiency standards for appliances and equipment that are covered by federal law under EPCA 42 U.S.C. 6201, et seq.[1] Thus, the IECC efficiencies related to mechanical systems are achieved with insulated and sealed delivery systems and controls.

The 2021 IECC Section R403 is divided into 12 parts:

- Sections R403.1 and R403.2 address controls for residential systems, including thermostats and hot water boiler outdoor temperature
- Section R403.3 covers ducts and air handlers
- Section R403.4 addresses insulation for system piping
- Section R403.5 addresses service hot water systems
- Section R403.6 covers mechanical ventilation provisions, including whole house fan efficiencies and system testing
- Section R403.7 addresses equipment sizing and efficiency rating
- Section R403.8 provides requirements for systems serving multiple dwelling units
- Section R403.9 addresses snow melt and ice system controls
- Sections R403.10 through R403.12 address pool and spa energy consumption.

These are the significant changes to the building mechanical systems requirements in the 2021 IECC:

- Clarified insulation requirements for ducts located outside of conditioned space, in conditioned space and buried
- Expanded duct testing requirements
- New mechanical ventilation system testing requirements

1 For purposes of this document, "EPCA" refers the EPCA as amended and codified in 42 U.S.C. §§ 6201-6422. Since its enactment in 1975, EPCA has been amended by the National Energy Conservation Policy Act, Pub. L. No. 95-619, § 422, 92 Stat. 3206 (1978); the National Appliance Energy Conservation Act, Pub. L. No. 100-12, § 7, 101 Stat. 103 (1987); the Energy Policy Act of 1992, Pub. L. 102-486, §§ 121-28, 106 Stat. 2776 (1992); the Energy Policy Act of 2005, Pub. L. 109-58, §§ 135-36, 119 Stat. 594 (2005); Energy Independence and Security Act of 2007,Pub. L. 110-140, §§ 301-325, 121 Stat. 1492 (2007); and the American Energy Manufacturing Technical Corrections Act,Pub. L. 112-210, 126 Stat. 1514 (2012).

R403.3.1-R403.3.3.1

Duct Location and Insulation

CHANGE TYPE: Modification

CHANGE SUMMARY: Ducts can be located in one of three locations: outside conditioned space, in conditioned space and buried. The 2021 IECC clarifies these locations and the insulation requirements for each.

2021 CODE: **R403.3.1 Ducts ~~Insulation~~ located outside conditioned space. ~~(Prescriptive).~~** Supply and return ducts located outside conditioned space shall be insulated to an R-value of not less than R-8 for ducts 3 inches (76 mm) in diameter and larger and not less than R-6 for ducts smaller than 3 inches (76 mm) in diameter. Ducts buried beneath a building shall be insulated as required per this section or have an equivalent thermal distribution efficiency. Underground ducts utilizing the thermal distribution efficiency method shall be listed and labeled to indicate the *R*-value equivalency.

R403.3.2 ~~R403.3.7~~ Ducts located in conditioned space. For duct~~s~~ work to be considered ~~as~~ inside a conditioned space, ~~such ducts~~ it shall comply with ~~either~~ one of the following:

1. The duct system shall be located completely within the continuous air barrier and within the building thermal envelope.
2. ~~The ducts shall~~ Ductwork in ventilated attic spaces shall be buried within ceiling insulation in accordance with Section ~~R403.3.6~~ R403.3.3 and all of the following conditions shall exist:
   - **2.1.** The air handler is located completely within the continuous air barrier and within the building thermal envelope.
   - **2.2.** The duct leakage, as measured either by a rough-in test of the ducts or a post-construction total system leakage test to outside the building thermal envelope in accordance with

Duct location determines insulation requirements.

Photo courtesy of BanksPhotos/iStock/Getty Images Plus

Section R403.3.6, is less than or equal to 1.5 cubic feet per minute (42.5 L/min) per 100 square feet (9.29 m^2) of conditioned floor area served by the duct system.

2.3. The ceiling insulation *R*-value installed against and above the insulated duct is greater than or equal to the proposed ceiling insulation *R*-value, less the *R*-value of the insulation on the duct.

3. Ductwork in floor cavities located over unconditioned space shall comply with all of the following:
   - **3.1.** A continuous air barrier installed between unconditioned space and the duct.
   - **3.2.** Insulation installed in accordance with Section R402.2.7.
   - **3.3.** A minimum R-19 insulation installed in the cavity width separating the duct from unconditioned space.
4. Ductwork located within exterior walls of the building thermal envelope shall comply with the following:
   - **4.1.** A continuous air barrier installed between unconditioned space and the duct.
   - **4.2.** Minimum R-10 insulation installed in the cavity width separating the duct from the outside sheathing.
   - **4.3.** The remainder of the cavity insulation shall be fully insulated to the drywall side.

R403.3.3 ~~R403.3.6~~ Ducts buried within ceiling insulation. Where supply and return air ducts are partially or completely buried in ceiling insulation, such ducts shall comply with all of the following:

1. The supply and return ducts shall have an insulation *R*-value not less than R-8.
2. At all points along each duct, the sum of the ceiling insulation *R*-value against and above the top of the duct, and against and below the bottom of the duct, shall be not less than R-19, excluding the *R*-value of the duct insulation.
3. In Climate Zones 0A, 1A, 2A and 3A, the supply ducts shall be completely buried within ceiling insulation, insulated to an R-value of not less than R-13 and in compliance with the vapor retarder requirements of Section 604.11 of the International Mechanical Code or Section M1601.4.6 of the International Residential Code, as applicable.

Exception: Sections of the supply duct that are less than 3 feet (914 mm) from the supply outlet shall not be required to comply with these requirements.

R403.3.3.1 ~~R403.3.6.1~~ Effective *R*-value of deeply buried ducts. Where using the Total Building Performance Compliance Option in accordance with Section R401.2.2 ~~simulated energy performance analysis~~, sections of ducts that are installed in accordance with Section R403.3.3, ~~R403.3.6~~ located directly on, or within 5.5 inches (140 mm) of the ceiling, surrounded with blown-in attic insulation having an *R*-value of R-30 or greater and located such that the top of the duct is not less than 3.5 inches (89 mm) below the top of the insulation, shall be considered as having an effective duct insulation *R*-value of R-25.

CHANGE SIGNIFICANCE: Section R403.3 includes both structural and technical changes. First, Section R403.3 is reorganized for clarity and application, making it more understandable and usable for field practitioners. From that reorganization, three distinct locations where ductwork can be installed—outside conditioned space, inside conditioned space and buried—are outlined along with the insulation requirements for each.

The 2018 IECC fully addressed buried duct provisions and updates to the 2021 IECC offer similar changes for ducts located within the continuous air barrier and building thermal envelope. Section R403.3.2 now lists conditions that must be met for ducts to be considered inside a conditioned space. Changes made to this section are intended to increase the energy performance of ductwork within homes.

Section R403.3.1 is applicable to all compliance paths except the Tropical Climate Region compliance option. Sections R403.3.2 and R403.3.3 are only applicable to the Prescriptive compliance option.

R403.3.5

Duct Testing

CHANGE TYPE: Modification

CHANGE SUMMARY: Ducts in conditioned spaces are no longer exempt from duct testing requirements and a duct testing standard is specified.

2021 CODE: **R403.3.5 ~~R403.3.3~~ Duct testing (~~Mandatory~~).** Ducts shall be pressure tested in accordance with ANSI/RESNET/ICC 380 or ASTM E1554 to determine air leakage by one of the following methods:

1. Rough-in test: Total leakage shall be measured with a pressure differential of 0.1 inch w.g. (25 Pa) across the system, including the manufacturer's air handler enclosure if installed at the time of the test. Registers shall be taped or otherwise sealed during the test.
2. Postconstruction test: Total leakage shall be measured with a pressure differential of 0.1 inch w.g. (25 Pa) across the entire system, including the manufacturer's air handler enclosure. Registers shall be taped or otherwise sealed during the test.

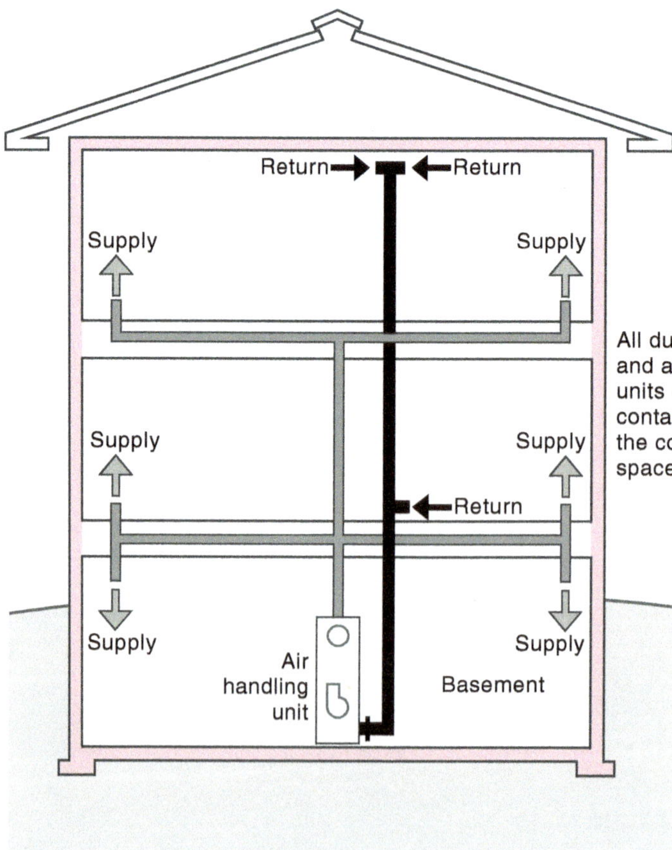

Graphic courtesy of Building America

Note: Colored shading depicts the building's thermal barrier and pressure boundary. The thermal barrier and pressure boundary enclose the conditioned space.

Ducts inside conditioned space must now be tested.

Exception:

1. ~~A duct air-leakage test shall not be required where the ducts and air handlers are located entirely within the building thermal envelope.~~
2. A duct air-leakage test shall not be required for ducts serving ventilation systems that are not integrated with ducts serving heating or cooling systems.

A written report of the results of the test shall be signed by the party conducting the test and provided to the code official.

CHANGE SIGNIFICANCE: Previously, the IECC exempted homes from duct testing requirements where the air handler and all ducts were located inside a conditioned space. Although moving all ducts inside a conditioned space may have a positive impact on energy efficiency overall, this practice alone cannot guarantee that the ducts will be tight enough to deliver conditioned air to all occupied areas of the home. Uncomfortable occupants commonly adjust thermostat settings to counteract the effect of poor delivery of conditioned air, leading to losses in energy efficiency. Additionally, homes with leaky ducts and poor air delivery are at far greater risk for builder callback. The 2021 IECC requires a duct leakage test for all new homes, including homes with all ducts inside conditioned space, mitigating the risk for such callbacks and ensuring ducts are functioning appropriately.

Duct leakage rates can be extremely high when ducts are not tested. Without an objective test as a means of quality assurance, even careful builders may not be aware of missed connections or poor sealing. In a recent DOE field study of residential homes in Kentucky, homes received duct leakage tests even where all supply and return ducts were located inside conditioned space.[2] Of the 24 homes tested with ducts inside conditioned spaces, the leakage rate averaged 18.5 cfm/ft^2, with individual homes ranging from 6.26 cfm/ft^2 to as high as 40.36 cfm/ft^2, higher leakage rates than the 2018 IECC requirement. Forty other homes in the same study with at least some ducts located outside the conditioned space achieved leakage rates of 9.7 cfm/ft^2, on average—roughly half the leakage rate of homes that qualified for the exemption. The field studies found similar results in Pennsylvania, where exempt homes (those with all ducts inside conditioned space) averaged almost 31 cfm/ft^2 leakage, while homes that were required to be tested averaged almost 18 cfm/ft^2 leakage.

The concept of requiring a test for all new homes is not new. DOE's Building America Program recommends that "[e]ven in conditioned space, ducts should be insulated to reduce the risk of condensation and mold. They should be tightly sealed and tested for leakage."[3]

This provision is applicable to all compliance paths except the Tropical Climate Region compliance option.

2 "Energy Efficiency Field Studies." *Building Energy Codes Program*, US Department of Energy. https://www.energycodes.gov/energy-efficiency-field-studies.

3 DOE Building Technologies Program. (January 2013). Ducts in Conditioned Space. US Department of Energy [PDF].

R403.6.3

Mechanical Ventilation System Testing

CHANGE TYPE: Addition

CHANGE SUMMARY: Mechanical ventilation systems must be tested to ensure they are installed and operating as intended.

2021 CODE: **R403.6.3 Testing.** Mechanical ventilation systems shall be tested and verified to provide the minimum ventilation flow rates required by Section R403.6. Testing shall be performed according to the ventilation equipment manufacturer's instructions, or by using a flow hood or box, flow grid, or other airflow measuring device at the mechanical ventilation fan's inlet terminals or grilles, outlet terminals or grilles, or in the connected ventilation ducts. Where required by the code official, testing shall be conducted by an approved third party. A written report of the results of the test shall be signed by the party conducting the test and provided to the code official.

> **Exception:** Kitchen range hoods that are ducted to the outside with 6-inch (152 mm) or larger duct and not more than one 90-degree (1.57 rad) elbow or equivalent in the duct run.

CHANGE SIGNIFICANCE: Minimum ventilation rates are indicated in R403.6. The testing experience gained through the verification of the ENERGY STAR program demonstrated that ventilation fans are installed but not performing as required by the code. Fan-rated flow does not equate

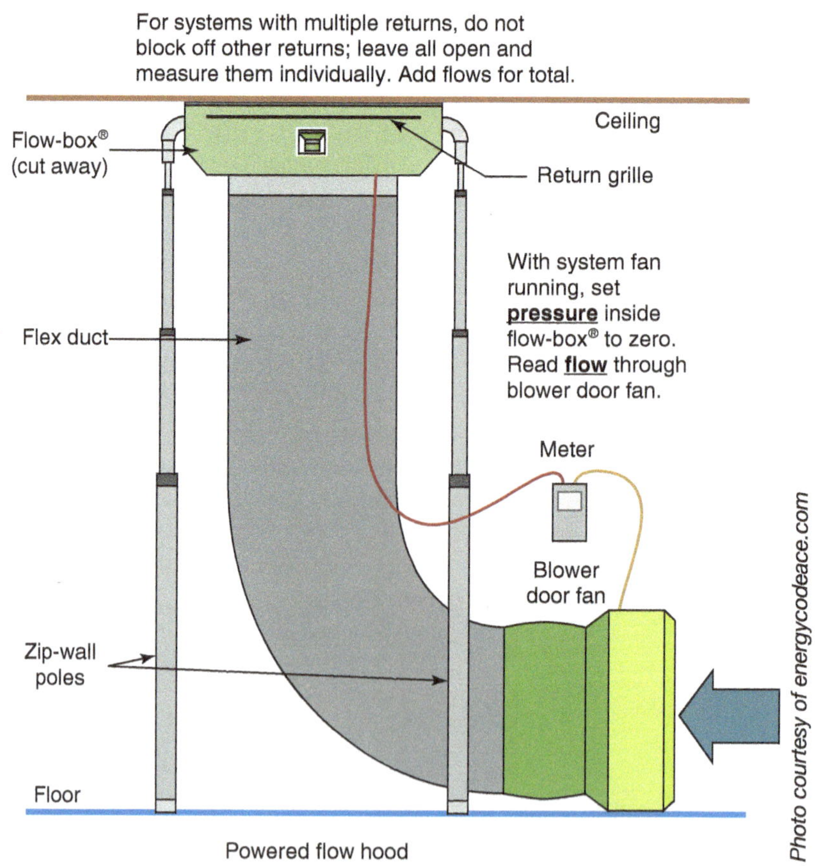

Mechanical ventilation system testing is required.

to the flow that is actually produced once a fan has been installed. The quality of the installation of the duct—from the fan, to the termination of the duct, to the outside—as well as the quality of the termination device ultimately governs the amount of air that any fan can push.

This new provision requires testing to ensure that the systems are installed in such a way that they work as intended by the code.

This provision is applicable to all compliance paths except the Tropical Climate Region compliance option.

R404

Electrical Power and Lighting Systems

Section R404 broadly addresses residential electrical power and lighting systems. Lighting uses about 10 percent of residential electrical power. Significant changes to the 2021 IECC electrical power and lighting requirements include both the introduction of exterior lighting requirements for some residential occupancies and controls into the residential provisions.

Photo courtesy of karamysh/iStock/Getty Images Plus

The 2021 IECC now regulates both interior and exterior lighting.

R404.1.1
Exterior Lighting

CHANGE TYPE: Addition

CHANGE SUMMARY: An exterior lighting requirement to comply with Section C405.5 is introduced for select residential buildings.

2021 CODE: **R404.1.1 Exterior lighting.** Connected exterior lighting for residential buildings shall comply with Section C405.5.

Exceptions:

1. Detached one- and two-family dwellings.
2. Townhouses.
3. Solar-powered lamps not connected to any electrical service.
4. Luminaires controlled by a motion sensor.
5. Lamps and luminaires that comply with Section R404.1.

CHANGE SIGNIFICANCE: Historically, the IECC has not included specific requirements for exterior lighting for residential buildings. The exterior lighting load may not be a significant issue for single-family homes, duplexes or townhomes, but it can be significant for Type-R occupancies because the load includes multiple-family parking lots and other exterior lighting applications.

The new Section R404.1.1 directs select residential occupancies to the commercial code and its lighting power density requirements. The section exempts detached one- and two-family residences, solar-powered lighting and any lighting controlled by a motion sensor.

When applied to the low-rise multiple-family prototype building developed by Pacific Northwest National Laboratory for the code determination studies, this requirement saved up to 0.5 percent (based on climate zone) of the whole building energy over the 2015 IECC. Since both the 2018 and 2015 IECC do not include exterior lighting requirements, this is a reasonable approximation of savings over the 2018 IECC as well.

This provision is applicable to all compliance paths.

Photo courtesy of rgaydos/E+/Getty Images

Exterior residential lighting must meet commercial requirements where applicable.

R404.2

Interior Lighting Controls

CHANGE TYPE: Addition

CHANGE SUMMARY: Permanently installed lighting fixtures are required to have either a dimmer, occupant sensor or other control unless exempt.

2021 CODE: **R404.2 Interior lighting controls.** Permanently installed lighting fixtures shall be controlled with either a dimmer, an occupant sensor control or other control that is installed or built into the fixture.

> **Exception:** Lighting controls shall not be required for the following:
>
> 1. Bathrooms.
> 2. Hallways.
> 3. Exterior lighting fixtures.
> 4. Lighting designed for safety or security.

CHANGE SIGNIFICANCE: Previous versions of the IECC have not included specific requirements for residential lighting controls. However, as has been proven with the commercial market, controls applied to efficient equipment and lighting can further efficiencies.

Section R404.2 requires interior lighting controls in the form of either a dimmer, occupancy control, switch or similar control, such as an automatic daylight sensor. Dimmers can reduce energy use by about 20 percent, while occupancy sensors reduce wasted energy by around 30 percent. These controls are essentially permanent due to their long lifetimes.

This provision is applicable to all compliance paths.

Photo courtesy of SDI Productions/E+/Getty Images

Interior lighting is controlled.

R404.3

Exterior Lighting Controls

CHANGE TYPE: Addition

CHANGE SUMMARY: Exterior lighting is required to have a manual on/off switch that permits automatic shutoff.

2021 CODE: **R404.3 Exterior lighting controls.** Where the total permanently installed exterior lighting power is greater than 30 watts, the permanently installed exterior lighting shall comply with the following:

1. Lighting shall be controlled by a manual on and off switch which permits automatic shut off actions.

Exception: Lighting serving multiple dwelling units.

2. Lighting shall be automatically shut off when daylight is present and satisfies the lighting needs.
3. Controls that override automatic shut off actions shall not be allowed unless the override automatically returns automatic control to its normal operation within 24 hours.

CHANGE SIGNIFICANCE: Similar to interior controls, Section R404.3 introduces residential exterior lighting provisions to the IECC. The new section requires controls for systems with greater than 30 watts; these controls must have manual switches that allow automatic shutoff and lighting must automatically shutoff based on daylight. The provisions also outline the limits to the override functions of the controls. The type of control is not specified, only what it must accomplish. The 2014 Consortium for Energy Efficiency report titled, "Residential Lighting Controls Market Characterization"[4] identifies that the use of a photosensor or timer can save, on average, 15 to 60KWh per year, based on the efficacy of the light source that is controlled.

This provision is applicable to the Prescriptive and Tropical Climate Region compliance options.

Photo courtesy of ShutterWorx/E+/Getty

Exterior lighting controls are required.

4 Consortium for Energy Efficiency. (January 9, 2014). Residential Lighting Controls Market Characterization [PDF]. https://library.cee1.org/system/files/library/11458/CEE_LightingMarketCharacterization.pdf.

R405

Total Building Performance

Section R405 contains the provisions for the Total Building Performance compliance option. This performance-based option offers flexibility in determining the most cost-effective methods to design a code-compliant structure. Total Building Performance uses software to predict the annual energy use of the proposed design compared to that of the standard design of building components listed in Table R405.4.2(1). If the proposed building's annual energy use is less than or equal to that of the standard design, the building complies with the code's goal of effective energy use. This method is useful to consider for any building designs that incorporate shading strategies to minimize air-conditioning loads in the summer and maximize heat gain in the winter.

The 2021 IECC includes a restructuring of the section to clarify required provisions (formerly referred to as mandatory) and updates to the standard reference design.

R405.2

Performance-Based Compliance

CHANGE TYPE: Addition

CHANGE SUMMARY: Table R405.2 lists required provisions for performance-based compliance.

2021 CODE: **R405.2 ~~Mandatory requirements~~ Performance-based compliance.** Compliance ~~with this section~~ based on total building performance requires that ~~the mandatory provisions identified in Section R401.2 be met. Supply and return ducts not completely inside the building thermal envelope shall be insulated to an R-value of not less than R-6.~~ a proposed design meets all of the following:

1. The requirements of the sections indicated within Table R405.2.
2. The building thermal envelope shall be greater than or equal to levels of efficiency and solar heat gain coefficients in Table R402.1.1 or R402.1.3 of the 2009 International Energy Conservation Code.
3. An annual energy cost that is less than or equal to the annual energy cost of the standard reference design. Energy prices shall be taken from a source approved by the code official, such as the Department of Energy, Energy Information Administration's State Energy Data System Prices and Expenditures reports. Code officials shall be permitted to require time-of-use pricing in energy cost calculations.

Exception: The energy use based on source energy expressed in Btu or Btu per square foot of conditioned floor area shall be permitted to be substituted for the energy cost. The source energy multiplier for electricity shall be 3.16. The source energy multiplier for fuels other than electricity shall be 1.1.

~~**R405.3 Performance-based compliance.** Compliance based on simulated energy performance requires that a proposed residence (proposed design) be shown to have an annual energy cost that is less than or equal to the annual energy cost of the standard reference design. Energy prices shall be taken from a source approved by the code official, such as the Department of Energy, Energy Information Administration's State Energy Data System Prices and Expenditures reports. Code officials shall be permitted to require time-of-use pricing in energy cost calculations.~~

~~**Exception:** The energy use based on source energy expressed in Btu or Btu per square foot of conditioned floor area shall be permitted to be substituted for the energy cost. The source energy multiplier for electricity shall be 3.16. The source energy multiplier for fuels other than electricity shall be 1.1.~~

TABLE R405.2 Requirements for Total Building Performance

Section[a]	Title
	General
R401.2.5	Additional energy efficiency
R401.3	Certificate
	Building Thermal Envelope
R402.1.1	Vapor retarder
R402.2.3	Eave baffle
R402.2.4.1	Access hatches and doors
R402.2.10.1	Crawl space wall insulation installations
R402.4.1.1	Installation
R402.4.1.2	Testing
R402.5	Maximum fenestration *U*-factor and SHGC
	Mechanical
R403.1	Controls
R403.3, including R403.3.1, except sections R403.3.2, R403.3.3 and R403.3.6	Ducts
R403.4	Mechanical system piping insulation
R403.5.1	Heated water circulation and temperature maintenance systems
R403.5.3	Drain water heat recovery units
R403.6	Mechanical ventilation
R403.7	Equipment sizing and efficiency rating
R403.8	Systems serving multiple dwelling units
R403.9	Snow melt and ice systems
R403.10	Energy consumption of pools and spas
R403.11	Portable spas
R403.12	Residential pools and permanent residential spas
	Electrical Power and Lighting Systems
R404.1	Lighting equipment
R404.2	Interior lighting controls

[a] Reference to a code section includes all the relative subsections except as indicated in the table.

CHANGE SIGNIFICANCE: The labels *prescriptive* and *mandatory* were introduced in the 2009 IECC. Although the terms were applied to various section and subsection titles throughout previous versions of the IECC, the terms were never defined and were sometimes used inconsistently.

Long-time code users generally understood that *mandatory* was intended to mean "non-tradeable" when using performance compliance options. That is, where the procedures or systems described within the section are included as part of the design, the requirements of that section must be met and cannot be traded off. *Prescriptive*, on the other hand, was intended to mean *mandatory* when using the prescriptive path but "tradeable" when using the performance path. However, this understanding is not explicitly based on code language, leading to confusion about the use of the terms.

The 2021 IECC eliminates the terms mandatory and prescriptive, and instead, all the requirements applicable to the R405 Total Building Performance are located within Section R405. Table R405.2 includes all the required provisions for this compliance path. Code users should also note the provision for additional energy efficiency included in section R401.2.5 requires projects complying with the Total Building Performance option to either include one item from Section R408 that is not included in the proposed design, or demonstrate the proposed design annual energy cost is less than or equal to 95 percent of the annual energy cost of the standard reference design.

This provision is only applicable to the Total Building Performance compliance option.

R405.3.2

Compliance Report

CHANGE TYPE: Modification

CHANGE SUMMARY: The performance path compliance report is simplified for greater usability and easier enforcement.

2021 CODE: **R405.3.2** ~~**R405. 4.2**~~ **Compliance report.** Compliance software tools shall generate a report that documents that the proposed design complies with Section R405.3. A compliance report on the proposed design shall be submitted with the application for the building permit. Upon completion of the building, a confirmed compliance report based on the ~~as-built~~ confirmed condition of the building shall be submitted to the code official before a certificate of occupancy is issued. ~~Batch sampling of buildings to determine energy code compliance shall only be allowed for stacked multiple-family units.~~ Compliance reports shall include information in accordance with Sections R405.3.2.1 and R405.3.2.2. ~~Where the proposed design of a building could be built on different sites where the cardinal orientation of the building on each site is different, compliance of the proposed design for the purposes of the application for the building permit shall be based on the worst-case orientation, worst-case configuration, worst-case building air leakage and worst- case duct leakage. Such worst-case parameters shall be used as inputs to the compliance software for energy analysis.~~

Performance path certificate requirements are expanded.

R405.3.2.1 ~~R405.4.2.1~~ Compliance report for permit application. A compliance report submitted with the application for building permit shall include the following:

1. Building street address, or other building site identification.
2. ~~A statement indicating that the proposed design complies with Section R405.3.~~ The name of the individual performing the analysis and generating the compliance report.
3. ~~An inspection checklist documenting the building component characteristics of the proposed design as indicated in Table R405.5.2(1). The inspection checklist shall show results for both the standard reference design and the proposed design with user inputs to the compliance software to generate the results.~~ The name and version of the compliance software tool.
4. ~~A site-specific energy analysis report that is in compliance with Section R405.3.~~ Documentation of all inputs entered into the software used to produce the results for the reference design and/or the rated home.
5. ~~The name of the individual performing the analysis and generating the report.~~ A certificate indicating that the proposed design complies with Section R405.3. The certificate shall document the building components' energy specifications that are included in the calculation including: component-level insulation *R*-values or *U*-factors; duct system and building envelope air leakage testing assumptions; and the type and rated efficiencies of proposed heating, cooling, mechanical ventilation and service water-heating equipment to be installed. If on-site renewable energy systems will be installed, the certificate shall report the type and production size of the proposed system.
6. ~~The name and version of the compliance software tool.~~ Where a site-specific report is not generated, the proposed design shall be based on the worst-case orientation and configuration of the rated home.

R405.3.2.2 ~~R405. 4.2.2~~ Compliance report for certificate of occupancy. A compliance report submitted for obtaining the certificate of occupancy shall include the following:

1. Building street address, or other building site identification.
2. Declaration of the total building ~~simulated~~ performance path on the title page of the energy report and the title page of the building plans.
3. A statement, bearing the name of the individual performing the analysis and generating the report, indicating that the as-built building complies with Section R405.3.
3. ~~A certificate indicating that the building passes the performance matrix for code compliance and indicating the energy saving features of the buildings.~~ The name and version of the compliance software tool.
4. A site-specific energy analysis report that is in compliance with Section R405.3.

5. ~~The name of the individual performing the analysis and generating the report.~~ A final confirmed certificate indicating compliance based on inspection, and a statement indicating that the confirmed rated design of the built home complies with Section R405.3. The certificate shall report the energy features that were confirmed to be in the home, including component-level insulation *R*-values or *U*-factors; results from any required duct system and building envelope air leakage testing; and the type and rated efficiencies of the heating, cooling, mechanical ventilation and service water-heating equipment installed.
6. ~~The name and version of the compliance software tool.~~ When on-site renewable energy systems have been installed, the certificate shall report the type and production size of the installed system.

~~**R405.4.3 Additional documentation.** The code official shall be permitted to require the following documents:~~

1. ~~Documentation of the building component characteristics of the standard reference design.~~
2. ~~A certification signed by the builder providing the building component characteristics of the proposed design as given in Table R405.5.2(1).~~
3. ~~Documentation of the actual values used in the software calculations for the proposed design.~~

CHANGE SIGNIFICANCE: The Total Building Performance compliance option has seen extensive use in Colorado, Florida and other parts of the country since the 2009 IECC, and these experiences continue to inform the language of the IECC. In the 2015 IECC, the process by which the Total Building Performance path was successfully implemented in Colorado was included in the code to help jurisdictions and builders better understand both how to implement the pathway to gain the flexibility of choice and tradeoff as well as how to verify compliance.

In the 2021 IECC, the Colorado experience further informs the code by simplifying the compliance report and clarifying documentation requirements. For example, the use of the terms *proposed* and *confirmed* to refer to documentation needed to obtain the building permit and certificate of occupancy is more clearly laid out. In addition, Sections R405.3.2.1 and R405.4.2.2 have been consolidated and reordered to more clearly identify for the jurisdiction having authority what documentation needs to be reviewed to demonstrate compliance.

These provisions are only applicable to the Total Building Performance compliance option.

R406

Energy Rating Index Compliance Alternative

Section R406 of the IECC contains the provisions for the Energy Rating Index (ERI) compliance option. Introduced in the 2015 IECC, the ERI is another path that can be used to demonstrate compliance with the energy code. The ERI is an index scoring system with a scale that ranges from 0 to 100. A net zero home or residential building will achieve an index value of 0; a home built to minimum requirements of the 2006 IECC is considered the reference design and has an index score of 100. Each point on the scale represents a 1-percent change in the total energy use of the rated design relative to the total energy use of the ERI reference design. The ERI score considers all energy used in the residential building.

Significant changes to the 2021 IECC include a restructuring of the section to clarify required provisions (formerly referred to as mandatory), updates to the target ERI scores and the introduction of provisions for renewable energy certificate documentation.

R406.2

ERI Compliance

CHANGE TYPE: Modification

CHANGE SUMMARY: Table R406.2 lists required provisions for the Energy Rating Index compliance option.

2021 CODE: **R406.2 ERI compliance.** Compliance based on the ERI requires that the rated design meets all of the following:

1. The requirements of the sections indicated within Table R406.2.
2. Maximum ERI of Table R406.5.

TABLE R406.2 **Requirements for Energy Rating Index**

Section[a]	Title
General	
R401.2.5	Additional efficiency packages
R401.3	Certificate
Building Thermal Envelope	
R402.1.1	Vapor retarder
R402.2.3	Eave baffle
R402.2.4.1	Access hatches and doors
R402.2.10.1	Crawl space wall insulation installation
R402.4.1.1	Installation
R402.4.1.2	Testing
Mechanical	
R403.1	Controls
R403.3, except sections R403.3.2, R403.3.3 and R403.3.6	Ducts
R403.4	Mechanical system piping insulation
R403.5.1	Heated water circulation and temperature maintenance systems
R403.5.3	Drain water heat recovery units
R403.6	Mechanical ventilation
R403.7	Equipment sizing and efficiency rating
R403.8	Systems serving multiple dwelling units
R403.9	Snow melt and ice systems
R403.10	Energy consumption of pools and spas
R403.11	Portable spas
R403.12	Residential pools and permanent residential spas
Electrical Power and Lighting Systems	
R404.1	Lighting equipment
R404.2	Interior lighting controls
R406.3	Building thermal envelope

[a]Reference to a code section includes all of the relative subsections except as indicated in the table.

CHANGE SIGNIFICANCE: As discussed throughout this book, the 2021 IECC eliminates the terms mandatory and prescriptive, and instead, all the requirements applicable to the R406 Energy Rating Index are located within Section R406. Table R406.2 outlines the required provisions for this compliance path.

Code users should also note the provision for additional energy efficiency included in section R401.2.5 requires projects complying with the ERI to demonstrate the ERI value is at least 5 percent less than the ERI target specified in Table R406.5.

This provision is only applicable to the Energy Rating Index compliance option.

R406.3

Building Thermal Envelope

CHANGE TYPE: Modification

CHANGE SUMMARY: Target ERI scores are lowered and the role of on-site renewables is clarified.

2021 CODE: **R406.3 Building thermal envelope.** Building and portions thereof shall comply with Section R406.3.1 or R406.3.2.

R406.3.1 On-site renewables are not included. Where on-site renewable energy is not included for compliance using the ERI analysis of Section R406.4, the proposed total building thermal envelope UA, which is sum of *U*-factor times assembly area, shall be less than or equal to the building thermal envelope UA using the prescriptive *U*-factors from Table R402.1.2 multiplied by 1.15 in accordance with Equation 4-1. The area-weighted maximum fenestration SHGC permitted in Climate Zones 0 through 3 shall be 0.30.~~levels of efficiency and Solar Heat Gain Coefficients in Table 402.1.1 or 402.1.3 of the 2009 International Energy Conservation Code.~~

$$UA_{\text{Proposed design}} = 1.15 \times UA_{\text{Prescriptive reference design}} \quad \textbf{(Equation 4-1)}$$

R406.3.2 On-site renewables are included. Where on-site renewable energy is included for compliance using the ERI analysis of Section R406.4, the building thermal envelope shall be greater than or equal to the levels of efficiency and SHGC in Table R402.1.2 or Table R402.1.4 of the 2018 International Energy Conservation Code.

TABLE R406.5 **Maximum Energy Rating Index**

CLIMATE ZONE	ENERGY RATING INDEX
0-1	~~57~~ 52
2	~~57~~ 52
3	~~57~~ 51
4	~~62~~ 54
5	~~61~~ 55
6	~~61~~ 54
7	~~58~~ 53
8	~~58~~ 53

~~a. Where on-site renewable energy is included for compliance using the ERI analysis of Section R406.4, the building shall meet the mandatory requirements of Section R406.2, and the building thermal envelope shall be greater than or equal to the levels of efficiency and SHGC in Table R402.1.2 or Table R402.1.4 of the 2015 *International Energy Conservation Code*.~~

CHANGE SIGNIFICANCE: Table R406.5 restores the lower ERI target scores from the 2015 IECC. Coupled with the requirements for additional energy efficiency in R401.2.5, which specify projects must reduce the ERI target by at least 5 percent, the 2021 ERI compliance option results in a significantly more efficient building than the 2018 IECC.

These provisions are only applicable to the Energy Rating Index compliance option.

R406.7.3

Renewable Energy Certificate Documentation

CHANGE TYPE: Addition

CHANGE SUMMARY: Renewable energy certificate (REC) documentation is introduced to ensure homeowners are able to legally document ownership of the RECs associated with their homes.

2021 CODE: **R406.7.3 Renewable energy certificate (REC) documentation.** Where on-site renewable energy is included in the calculation of an ERI, one of the following forms of documentation shall be provided to the code official:

1. Substantiation that the RECs associated with the on-site renewable energy are owned by, or retired on behalf of, the homeowner.
2. A contract that conveys to the homeowner the RECs associated with the on-site renewable energy, or conveys to the homeowner an equivalent quantity of RECs associated with other renewable energy.

CHANGE SIGNIFICANCE: Renewable energy has been an aspect of the IECC since its inception in 1998. However, as the industry has evolved in complexity, so have the issues of documentation and attribution.

The new section R406.7.3 addresses ownership of the environmental attributes of an on-site renewable energy system. The environmental attributes of solar power and other renewable energy systems have a market value that is reflected and transacted in RECs. This provision ensures the attributes are not double-counted by other entities by requiring legal documentation of ownership.

This section is supported by a new definition of renewable energy certificate that provides clarification on what the renewable energy certificate represents.

This provision is only applicable to the Energy Rating Index compliance option.

RECs represent the environmental attributes of renewable energy systems.

R407

Tropical Climate Region

CHANGE TYPE: Modification

CHANGE SUMMARY: The Tropical Region Compliance option is now located in Section R407.

2021 CODE:

SECTION R407
TROPICAL CLIMATE REGION COMPLIANCE PATH

R407.1 Scope. This section establishes alternative criteria for residential buildings in the tropical region at elevations less than 2,400 feet (731.5 m) above sea level.

R407.2 Tropical climate region. ~~**R401.2.1 Tropical zone.** Residential buildings in the tropical zone at elevations less than 2,400 feet (731.5 m) above sea level shall be deemed to be in compliance with this chapter provided that the following conditions are met~~ Compliance with this sections requires the following:

1. Not more than one-half of the occupied space is air conditioned.
2. The occupied space is not heated.
3. Solar, wind or other renewable energy source supplies not less than 80 percent of the energy for service water heating.
4. Glazing in conditioned spaces has a solar heat gain coefficient (SHGC) of less than or equal to 0.40, or has an overhang with a projection factor equal to or greater than 0.30.
5. Permanently installed lighting is in accordance with Section R404.
6. The exterior roof surface complies with one of the options in Table C402.3 of the International Energy Conservation Code–Commercial Provisions or the roof or ceiling has insulation with

Photo courtesy of Brand X Pictures/Stockbyte/Getty Images

Dwellings in the tropical zone may comply with the Tropical Climate Region option.

an *R*-value of R-15 or greater. Where attics are present, attics above the insulation are vented and attics below the insulation are unvented.

7. Roof surfaces have a slope of not less than 1/4 unit vertical in 12 units horizontal (21-percent slope). The finished roof does not have water accumulation areas.
8. Operable fenestration provides a ventilation area of not less than 14 percent of the floor area in each room. Alternatively, equivalent ventilation is provided by a ventilation fan.
9. Bedrooms with exterior walls facing two different directions have operable fenestration on exterior walls facing two directions.
10. Interior doors to bedrooms are capable of being secured in the open position.
11. A ceiling fan or ceiling fan rough-in is provided for bedrooms and the largest space that is not used as a bedroom.

CHANGE SIGNIFICANCE: The provisions for compliance via the alternative Tropical Climate Region Compliance path were originally located in Section R401.2.1, and there was some confusion as to the application of the compliance path. The creation of Section R407 aligns with the modifications to Section R401.2, clearly and consistently identifying each compliance path and providing each compliance path its own section. No technical changes have been made.

These provisions are only applicable to the Tropical Climate Region compliance option.

R408

Additional Efficiency Package Options

CHANGE TYPE: Addition

CHANGE SUMMARY: Residential projects are required to include additional energy efficiency measures as specified by the compliance path option selected. Section R408 outlines the additional energy efficiency package options for the Prescriptive and Total Building Performance paths.

2021 CODE:

SECTION R408
ADDITIONAL EFFICIENCY PACKAGE OPTIONS

R408.1 Scope. This section establishes additional efficiency package options to achieve additional energy efficiency in accordance with Section R401.2.5.

More efficient HVAC equipment is one additional efficiency package option.

R408.2 Additional efficiency package options. Additional efficiency package options for compliance with Section R401.2.1 are set forth in Sections R408.2.1 through R408.2.5.

R408.2.1 Enhanced envelope performance option. The total building thermal envelope UA, the sum of *U*-factor times assembly area, shall be less than or equal to 95 percent of the total UA resulting from multiplying the *U*-factors in Table R402.1.2 by the same assembly area as in the proposed building. The UA calculation shall be performed in accordance with Section R402.1.5. The area-weighted average SHGC of all glazed fenestration shall be less than or equal to 95 percent of the maximum glazed fenestration SHGC in Table R402.1.2.

R408.2.2 More efficient HVAC equipment performance option. Heating and cooling equipment shall meet or exceed one of the following efficiencies:

1. Greater than or equal to 95 AFUE natural gas furnace and 16 SEER air conditioner.
2. Greater than or equal to 10 HSPF/16 SEER air source heat pump.
3. Greater than or equal to 3.5 COP ground source heat pump.
4. For multiple cooling systems, all systems shall meet or exceed the minimum efficiency requirements in this section and shall be sized to serve 100 percent of the cooling design load. For multiple heating systems, all systems shall meet or exceed the minimum efficiency requirements in this section and shall be sized to serve 100 percent of the heating design load.

R408.2.3 Reduced energy use in service water-heating option. The hot water system shall meet or exceed one of the following efficiencies:

1. Greater than or equal to 82 EF fossil fuel service water-heating system.
2. Greater than or equal to 2.0 EF electric service water-heating system.
3. Greater than or equal to 0.4 Solar Fraction solar water-heating system.

R408.2.4 More efficient duct thermal distribution system option. The thermal distribution system shall meet or exceed one of the following efficiencies:

1. 100 percent of ducts and air handlers located entirely within the building thermal envelope.
2. 100 percent of ductless thermal distribution system or hydronic thermal distribution system located completely inside the building thermal envelope.
3. 100 percent of duct thermal distribution system located in conditioned space as defined by R403.3.2.

R408.2.5 Improved air sealing and efficient ventilation system option. The measured air leakage rate shall be less than or equal to 3.0 ACH50, with either an Energy Recovery Ventilator (ERV) or Heat Recovery Ventilator (HRV) installed. Minimum HRV and ERV requirements, measured at the lowest tested net supply airflow, shall be greater than or equal to 75 percent Sensible Recovery Efficiency (SRE), less than or equal to 1.1 cubic feet per minute per wall (0.03m^3/min/watt) and shall not use recirculation as a defrost strategy. In addition, the ERV shall be greater than or equal to 50 percent Latent Recovery/Moisture Transfer (LRMT).

CHANGE SIGNIFICANCE: Arguably the most significant change to the residential provisions of the 2021 IECC, Section R408 provides five options from which residential projects can choose to demonstrate additional energy efficiency. These improvements increase energy savings and reduce costs to the homeowner over the useful life of the building. Section R408 largely mirrors the format of the additional efficiency package options of the commercial provisions (Section C406) as included in the 2012 through the 2018 IECC, and is modeled after code language adopted in states like Oregon and Washington that have successfully created a list of options available to builders to meet the residential code improvements. This approach is intended to increase flexibility for code users while advancing the code's efficiency baseline.

Projects complying with the residential energy code prescriptively must include one of these packages from R408: enhanced envelope performance, more efficient HVAC equipment, reduced energy use in service water-heating, more efficient duct thermal distribution system or improved air sealing and efficiency ventilation system.

Projects using the Total Building Performance compliance option must either incorporate one package into the project without including it in the proposed design, or the proposed design must have an annual energy cost less than or equal to 95 percent of the standard reference design. These package options do not apply to the Energy Rating Index or Tropical Climate Region compliance options.

PART 8

Residential Energy Provisions: Residential Appendices

- **Appendix RA** Board of Appeals – Residential
- **Appendix RB** Solar-Ready Provisions – Detached One- and Two-Family Dwellings and Townhouses
- **Appendix RC** Zero Energy Residential Building Provisions

Appendix RA

Board of Appeals – Residential

RB 103.5

Shading

Appendix RC

Zero Energy Residential Building Provisions

Appendices to the *International Energy Conservation Code* (IECC) provide options for jurisdictions to expand the requirements of the IECC. These options are not mandatory unless specifically referenced in the adopting ordinances. The 2021 IECC introduces two new residential appendices – Appendix RA provides guidance for establishing a board of appeals and Appendix RC outlines provisions for zero energy residential buildings. The existing solar-ready appendix is also modified. ■

Appendix RA

Board of Appeals–Residential

CHANGE TYPE: Addition

CHANGE SUMMARY: A new Appendix RA provides guidance for establishing a board of appeals, including criteria for membership and instruction for developing rules and procedures.

2021 CODE:

APPENDIX RA
BOARD OF APPEALS—RESIDENTIAL

SECTION RA101
GENERAL

RA101.1 Scope. A board of appeals shall be established within the jurisdiction for the purpose of hearing applications for modification of the requirements of this code pursuant to the provisions of Section R110. The board shall be established and operated in accordance with this section, and shall be authorized to hear evidence from appellants and the code official pertaining to the application and intent of this code for the purpose of issuing orders pursuant to these provisions.

RA101.2 Application for appeal. Any person shall have the right to appeal a decision of the code official to the board. An application for appeal shall be based on a claim that the intent of this code or the rules legally adopted hereunder have been incorrectly interpreted, the provisions of this code do not fully apply or an equally good or better form of construction is proposed. The application shall be filed on a form obtained from the code official within 20 days after the notice was served.

Photo courtesy of SDI Productions/iStock/Getty Images Plus

Any person may appeal a decision of the code official.

RA101.2.1 Limitation of authority. The board shall not have authority to waive requirements of this code or interpret the administration of this code.

RA101.2.2 Stays of enforcement. Appeals of notice and orders, other than Imminent Danger notices, shall stay the enforcement of the notice and order until the appeal is heard by the board.

RA101.3 Membership of board. The board shall consist of five voting members appointed by the chief appointing authority of the jurisdiction. Each member shall serve for [INSERT NUMBER OF YEARS] years or until a successor has been appointed. The board members' terms shall be staggered at intervals, so as to provide continuity. The code official shall be an ex officio member of said board but shall not vote on any matter before the board.

RA101.3.1 Qualifications. The board shall consist of five individuals, who are qualified by experience and training to pass on matters pertaining to building construction and are not employees of the jurisdiction.

RA101.3.2 Alternate members. The chief appointing authority is authorized to appoint two alternate members who shall be called by the board chairperson to hear appeals during the absence or disqualification of a member. Alternate members shall possess the qualifications required for board membership, and shall be appointed for the same term or until a successor has been appointed.

RA101.3.3 Vacancies. Vacancies shall be filled for an unexpired term in the same manner in which original appointments are required to be made.

RA101.3.4 Chairperson. The board shall annually select one of its members to serve as chairperson.

RA101.3.5 Secretary. The chief appointing authority shall designate a qualified clerk to serve as secretary to the board. The secretary shall file a detailed record of all proceedings which shall set forth the reasons for the board's decision, the vote of each member, the absence of a member and any failure of a member to vote.

RA101.3.6 Conflict of interest. A member with any personal, professional or financial interest in a matter before the board shall declare such interest and refrain from participating in discussions, deliberations and voting on such matters.

RA101.3.7 Compensation of members. Compensation of members shall be determined by law.

RA101.3.8 Removal from the board. A member shall be removed from the board prior to the end of their terms only for cause. Any member with continued absence from regular meeting of the board may be removed at the discretion of the chief appointing authority.

RA101.4 Rules and procedures. The board shall establish policies and procedures necessary to carry out its duties consistent with the provisions of this code and applicable state law. The procedures shall not require compliance with strict rules of evidence, but shall mandate that only relevant information be presented.

RA101.5 Notice of meeting. The board shall meet upon notice from the chairperson, within 10 days of the filing of an appeal or at stated periodic intervals.

RA101.5.1 Open hearing. All hearings before the board shall be open to the public. The appellant, the appellant's representative, the code official and any person whose interests are affected shall be given an opportunity to be heard.

RA101.5.2 Quorum. Three members of the board shall constitute a quorum.

RA101.5.3 Postponed hearing. When five members are not present to hear an appeal, either the appellant or the appellant's representative shall have the right to request a postponement of the hearing.

RA101.6 Legal counsel. The jurisdiction shall furnish legal counsel to the board to provide members with general legal advice concerning matters before them for consideration. Members shall be represented by legal counsel at the jurisdiction's expense in all matters arising from service within the scope of their duties.

RA101.7 Board decision. The board shall only modify or reverse the decision of the code official by a concurring vote of three or more members.

RA101.7.1 Resolution. The decision of the board shall be by resolution. Every decision shall be promptly filed in writing in the office of the code official within three days and shall be open to the public for inspection. A certified copy shall be furnished to the appellant or the appellant's representative and to the code official.

RA101.7.2 Administration. The code official shall take immediate action in accordance with the decision of the board.

RA101.8 Court review. Any person, whether or not a previous party of the appeal, shall have the right to apply to the appropriate court for a writ of certiorari to correct errors of law. Application for review shall be made in the manner and time required by law following the filing of the decision in the office of the chief administrative officer.

CHANGE SIGNIFICANCE: Appendix RA mirrors the requirements of the 2018 *International Building Code* (IBC), 2018 *International Fire Code* (IFC) and 2021 *International Residential Code* (IRC), providing a consistent set of requirements for a board of appeals across the family of I-Codes. This appendix offers a framework for jurisdictions to establish a functional board of appeals, outlining the rights and responsibilities of the appellant, code official and board members. The provisions contained in this appendix are not mandatory unless specified as such in the jurisdiction's adopting ordinance.

RB103.5

Shading

CHANGE TYPE: Addition

CHANGE SUMMARY: To maximize solar technology effectiveness, Appendix RB defines a minimum setback for the designated solar-ready zone from permanently affixed objects and focuses on roof slopes facing sunward with sufficient slope to maximize energy collection.

2021 CODE: **RB103.5 Shading.** The solar-ready zone shall be set back from any existing or new permanently affixed object on the building or site that is located south, east or west of the solar zone a distance not less than two times the object's height above the nearest point on the roof surface. Such objects include, but are not limited to, taller portions of the building itself, parapets, chimneys, antennas, signage, rooftop equipment, trees and roof plantings.

RB103.6 Capped roof penetration sleeve. A capped roof penetration sleeve shall be provided adjacent to a solar-ready zone located on a roof slope of not greater than 1 unit vertical in 12 units horizontal (8-percent slope). The capped roof penetration sleeve shall be sized to accommodate the future photovoltaic system conduit but shall have an inside diameter of not less than 1 1/4 inches (32 mm).

CHANGE SIGNIFICANCE: The Solar-Ready Zone appendix (Appendix T) to the 2018 *International Residential Code* (IRC) was updated to include shading and capped roof penetration sleeve requirements; however, these updates were not incorporated into the Solar-Ready Provisions appendix to the 2018 IECC. These new code provisions not only better align the appendices of the 2021 IRC and 2021 IECC, but also provide clarification for and strengthen the existing Solar-Ready Provisions appendix. Section RB103.5 clarifies the concept of "shading" by defining the setback for the

Photo courtesy of Westend61/Getty Images

Roof with installed solar panels meeting the new shading requirements.

designated solar-ready zone from permanently affixed objects. Section RB103.6 requires a sleeve to provide a path for future wiring from the solar panel to the meter at the time of new construction; as such, roofs with a slope of 1:12 or less must be provided with a pipe sleeve penetration. For roofs with slopes greater than 1:12, other design options are available and a penetration is not necessary.

The Solar-Ready Provisions appendix of the IECC was also renumbered from Appendix RA to Appendix RB. The provisions contained in this appendix are not mandatory unless specified as such in the jurisdiction's adopting ordinance.

Appendix RC

Zero Energy Residential Building Provisions

CHANGE TYPE: Addition

CHANGE SUMMARY: Appendix RC provides a model for residential construction to achieve net zero energy consumption through the use of renewable energy systems.

2021 CODE:

APPENDIX RC
ZERO ENERGY RESIDENTIAL BUILDING PROVISIONS

SECTION RC101
COMPLIANCE

RC101.1 Compliance. Existing residential buildings shall comply with Chapter 5. New residential buildings shall comply with Section RC102.

SECTION RC102
ZERO ENERGY RESIDENTIAL BUILDINGS

RC102.1 General. New residential buildings shall comply with Section RC102.2.

RC102.2 Energy Rating Index zero energy score. Compliance with this section requires that the rated design be shown to have a score less than or equal to the values in Table RC102.2 when compared to the Energy Rating Index (ERI) reference design determined in accordance with RESNET/ICC 301 for both of the following:

1. ERI value not including on-site power production (OPP) calculated in accordance with RESNET/ICC 301.

Photo courtesy of jhorrocks/iStock/Getty Images Plus

On-site solar for net zero energy consumption.

2. ERI value including on-site power production calculated in accordance with RESNET/ICC 301 with the OPP in Equation 4.1.2 of RESNET/ICC 301 adjusted in accordance with Equation RC-1.

   Adjusted OPP = OPP + CREF + REPC (Equation RC-1)

   where:

   CREF = Community Renewable Energy Facility power production—the yearly energy, in kilowatt hour equivalent (kWheq), contracted from a community renewable energy facility that is qualified under applicable state and local utility statutes and rules, and that allocates bill credits to the rated home.

   REPC = Renewable Energy Purchase Contract power production—the yearly energy, in kilowatt hour equivalent (kWheq), contracted from an energy facility that generates energy with photovoltaic, solar thermal, geothermal energy or wind systems, and that is demonstrated by an energy purchase contract or lease with a duration of not less than 15 years.

TABLE RC102.2 Maximum Energy Rating Index[a]

CLIMATE ZONE	ENERGY RATING INDEX not including OPP	ENERGY RATING INDEX including Adjusted OPP (as proposed)
1	43	0
2	45	0
3	47	0
4	47	0
5	47	0
6	46	0
7	46	0
8	46	0

a. The building shall meet the requirements of Table R406.2, and the building thermal envelope shall be greater than or equal to the levels of efficiency and SHGC in Table R402.1.2 or Table RR402.1.3 of the 2015 International Energy Conservation Code.

CHANGE SIGNIFICANCE: Appendix RC provides adoptable language that is easily incorporated into the 2021 IECC and results in residential buildings that have zero energy consumption over the course of a year. The appendix builds on the familiar Energy Rating Index compliance option of the IECC (Section R406) by requiring the use of RESNET/ICC 301. This standard compares the energy performance of a home to a reference home of the same geometry modeled to the 2006 IECC; where the home requires no net purchased energy, the ERI is 0 and the home is considered to have net zero energy consumption.

Appendix RC requires projects to meet or exceed two separate ERI targets. First, projects must achieve a target ERI of 43 to 47, depending on climate zone, without considering on-site power production. The project must then account for renewable energy and achieve an ERI of 0.

Renewable energy may be a combination of on-site power production, energy generated through community renewable energy facilities and renewable energy purchase contracts or leases.

Jurisdictions may choose to adopt this appendix in support of policy goals related to energy efficiency and renewable energy. The provisions contained in this appendix are not mandatory unless specified as such in the jurisdiction's adopting ordinance.

ICC INTERNATIONAL CODE COUNCIL®

People Helping People Build a Safer World®

Proctored Remote Online Testing Option (PRONTO™)

Convenient, Reliable, and Secure Certification Exams

Take the Test at Your Location

Take advantage of ICC PRONTO, an industry leading, secure online exam delivery service. PRONTO allows you to take ICC Certification exams at your convenience in the privacy of your own home, office or other secure location. Plus, you'll know your pass/fail status immediately upon completion.

- With PRONTO, ICC's Proctored Remote Online Testing Option, take your ICC Certification exam from any location with high-speed internet access.
- With online proctoring and exam security features you can be confident in the integrity of the testing process and exam results.,
- Plan your exam for the day and time most convenient for you. PRONTO is available 24/7.
- Eliminate the waiting period and get your results in private immediately upon exam completion.
- #1 ICC was the first model code organization to offer secured online proctored exams—part of our commitment to offering the latest technology-based solutions to help building and code professionals succeed and advance. We continue to expand our catalog of PRONTO exam offerings.

Discover ICC PRONTO and the wealth of certification opportunities available to advance your career: www.iccsafe.org/MeetPRONTO

ANSI ACCREDITED
ISO/IEC 17065
Product Certification Body
#1000

ASSESSMENT center

pronto
ICC ONLINE EXAMS

20-19180